Contributions to Management Science

For further volumes:
http://www.springer.com/series/1505

Lalit Wankhade • Balaji Dabade

Quality Uncertainty and Perception

Information Asymmetry and Management of Quality Uncertainty and Quality Perception

Physica-Verlag

Lalit Wankhade
SGGS Institute of
Engineering & Technology
Dept. Production Engineering
431606 Vishnupuri
Nanded, Maharashtra
India
lalitwankhade@gmail.com

Prof. Dr. Balaji Dabade
SGGS Institute of
Engineering & Technology
Dept. Production Engineering
431606 Vishnupuri
Nanded, Maharashtra
India
bmdabade@gmail.com

ISBN 978-3-7908-2822-1 ISBN 978-3-7908-2195-6(eBook)
DOI 10.1007/978-3-7908-2195-6
Springer Heidelberg Dordrecht London New York

Cover design: WMXDesign GmbH, Heidelberg, Germany

printed on acid-free paper

Physica-Verlag is a brand of Springer-Verlag Berlin Heidelberg
Springer-Verlag is a part of Springer Science+Business Media (www.springer.com)

Preface

It has been observed that the studies of quality are pursued in various disciplines like economics, quality management, and marketing science, and are seen isolated. The treatments imparted to these studies are also different and has the backdrop of discipline in which the work has been pursued. The nature of isolation is equally seen when quality uncertainty and perceived quality were pursued separately without showing any inkling that these can be complimentary.

Economist and Nobel Laureate, Akerlof (1970), wrote a seminal piece "The market for lemons: quality uncertainty and market mechanism", where he described quality uncertainty due to information asymmetry. It refers to the fact that a party in a transaction may have more information than the other. This is information asymmetry. If the seller has more information than the buyer about the product quality, he/she may sell it, as if it is a high-quality product. In reality, it could be a low-quality product. The buyer does not have the information regarding the quality of the offered product. The market condition that led to this transaction is quality uncertainty due to information asymmetry.

An attempt was made in 1980s by Parasuraman, Zeithaml, and Berry, to develop a metric for the measurement of perceived quality. However, it was largely applicable to service quality. The approach to perceived quality of products has been a peculiar marketing science way of statistical computation on the basis of sample survey. It has neglected the Total Quality Management (TQM) efforts and the crucial component of information asymmetry that prevails in the market.

Traces of perceived quality are also seen in quality management. Garvin (1984) identified the eight dimensions of quality. Perceived quality is one of these eight dimensions. Garvin suggested that information content is one of the bases of perceived quality. The process of quality perception requires the basis of information that is similar to the phenomenon of quality uncertainty.

Thus, the dimensions of quality uncertainty and perceived quality have been studied in various disciplines. Now, it is a time to break the barriers of disciplines to usher in a more purposeful study on these aspects of quality. A few more disciplines can be involved to further the multidisciplinary approach. Uncertainties are handled by using probabilities. Hence, perception of quality is now probabilistically coined as quality perception as an opposite of quality uncertainty. Quality perception and quality uncertainty have become two sides of the same coin.

Theories of reliability engineering can be applied for modeling and analysis of quality uncertainty as well as quality perception. Fault tree and success tree methods are specifically helpful. Thus, economics, marketing science, reliability engineering, and quality management are mingled to evolve the theory of quality perception. The theories need to be refined and implemented by using system behavior approach. System dynamics then plays a final role in deciphering the complex situation of quality perception.

Intended Audience

At the core, the content of this book refers to economics, marketing, and quality management. Hence the audience for this book are primarily drawn from all these streams. Specifically, quality management people may want to incorporate the new dimension of quality perception that will render the competitive edge to business growth. The people from economics of information would like to see the way quality uncertainty issue is modeled and how simulations are useful to reiterate on hypothesis. Above all, it is marketing that has to deliver truly to the customers. The kernel of business growth lies in understanding the marketing strategies and the way strategies are developed. The work in this book supports to develop marketing strategies by using optimal resources, and also tries to directly link these strategies to business growth.

Also, the students from business schools can learn the engineering based mechanics of quality perception that is developed in this book. Many times, the management tools alone are not sufficient to fully explore the issue that is being handled. Quality uncertainty and perception have been given the multidisciplinary orientation in this book that is based on engineering perspective.

Developing the metric linking customers to business growth has been the constant urge of practicing executives and managers. They have implemented such metrics that were developed earlier, and the satisfaction had been either partial or none. Certainly, the work in this book will provide the scientific way of looking at quality perception which will render the proper insight to executives and managers. The insight gained from this book will be capable of developing managerial perspective to translate the customers' sentiments into business performance.

Organization of the Book

Quality uncertainty has been studied earlier in economics of information. These attempts are reviewed as an introduction to this book. The work starts by identifying the role of information asymmetry to understand the phenomenon of quality uncertainty. The mechanics are elaborated in the second chapter. Similarly, perceived quality as noticed in the literature is presented in the third chapter. Some attempts have been made earlier, to develop the metric for measurement of perceived quality, which are traced through the time and disciplines.

The fourth chapter links quality perception to quality uncertainty. The extent of information asymmetry or symmetry has been used as the basis for this linkage. This is a chapter from where the disciplines start mingling in this book. Quality uncertainty or perception is viewed from both, endogenous and exogenous perspectives. Total Quality Management makes the endogenous side with TQM constructs as endogenous variables, whereas marketing science framework exhibits the exogenous side. Principles of economics of information are implemented to correlate marketing parameters with the TQM practices in industry so that quality is perceived by customers. Quality uncertainty and perception behaviors are also related to product life cycle. Thus, a new direction has been given to the perceived quality that traversed thus far in different streams.

Generally data collection is a tedious work discouraging managers from implementing any theoretical framework. The approximate method that creates starting motivation is proposed in the fifth chapter. It needs only partial data and identification of the product specific factors. Affinity diagram, interrelationship diagram, and tree diagram, from seven management tools are used to develop root cause and failure analysis of quality uncertainty.

The sixth chapter presents the system dynamics approach to modeling and analysis of quality perception. Models are developed in Vensim PLE. Simulation runs are carried out. The process of quality perception is made explicit through simulations. Behaviors are studied extensively. Thus, the dynamics of quality perception are presented in this book. Finally, research potential on the topic is unfolded in the seventh chapter.

The instruments to minimize quality uncertainty or to maximize quality perception are developed in the fourth, fifth and sixth chapter. The instruments are developed along with subtle explanation so that their use in real life should be credible. The reader tends to get stuck in all these details and using these tools look difficult. The seeming complexity of using the tools has been avoided by developing the simple worksheets that are ready for practical use. Thus, six worksheets are developed and included in appendices at the end of the relevant chapters. The step-by-step explanation is also provided to fill the worksheets.

Nanded, India

Lalit Wankhade
Balaji Dabade

Acknowledgement

Producing a book is a complex undertaking, and it involves the work of many people. The authors and the publishers are thankful to Emerald Group Publishing Limited for granting permission to use following research papers for this book.

Wankhade L, Dabade BM (2005) A holistic approach to quality success in developing nations. TQM Mag 17(4):322–328

Wankhade L, Dabade BM (2006) Analysis of quality uncertainty due to information asymmetry. Int J Qual Reliab Manage 23(2):230–241

Wankhade L, Dabade BM (2006) TQM with quality perception: a system dynamics approach. TQM Mag 18(4):341–357

Wankhade L, Dabade BM (2007) Information symmetry, quality perception, and market dynamics. J Model Manage 2(3):208–231

The authors and the publishers are also thankful to Inderscience Enterprises Limited for granting permission to use the following research papers for this book.

Wankhade L, Dabade BM (2009) Minimising quality uncertainty by the root-cause and failure analysis method. Int J Manage Decis Mak 10(5/6):359–385

The authors are grateful to all the reviewers who offered valuable suggestions during the research work and book writing. The authors are thankful to Christian Rauscher and the Springer-Verlag GmbH team, and truly appreciate their support extended towards the development of this book.

Finally, the first author thanks his wife Kavita and son Tanmay for their love, patience, and support. The second author thanks his wife Sushama and son Sanket for their love and support.

Contents

Chapter 1
Introduction

Defining quality is a stupendous task in itself. A wide range of definitions have been offered in the course of its evolution. Product acceptance or rejection by inspection was the early stage of quality control. Finding the inspection of each product a cumbersome procedure, sampling methods were devised. This, along with statistical tools, gave rise to statistical quality control. In course of time, various new techniques and methods followed. A few definitions of quality are given below:

"Quality consists of the capacity to satisfy wants." (C.D. Edwards, "The Meaning of Quality," in *Quality Progress*, Oct. 1968)

"Quality is fitness for use." (J.M. Juran, ed. *Quality Control Handbook*, 1988)

"Quality is the degree to which a specific product conforms to a design or specification" (H.L. Gilmore: "Product Conformance Cost," in *Quality progress*, June 1974)

"Quality [means] conformance to requirements." (P.B. Crosby: Quality Is Free)

"Quality refers to the amount of the unpriced attributes contained in each unit of the priced attribute." (K. B. Leifler: "Ambiguous Changes in Product Quality," *American Economic Review*, Dec. 1982)

"Quality is the degree of excellence at an acceptable price and the control of variability at an acceptable cost." (R. A. Broh: "Managing Quality for Higher Profits," 1982)

"Quality is a customer determination based upon a customer's actual experience with a product or service, measured against his or her requirements – stated or unstated, conscious or merely sensed, technically operational or entirely subjective – and always representing a moving target in a competitive market." (Armand Feigenbaum)

"Quality is a transformation in the way we think and work together, in what we value and reward, and in the way we measure success. All of us collaborate to design and operate a seamless value-adding system that incorporates quality control, customer service, process improvement, supplier relationships, and good relations with the communities we serve and in which we operate – all optimizing for a common purpose." (Peter Senge et al. *The Fifth Discipline Field book*)

"Quality denotes an excellence in goods and services, especially to the degree they conform to requirements and satisfy customers." (American Society for Quality)

Many quality experts, Deming (1981, 1982, 1986), Juran (1974, 1986, 1988), Crosby (1979, 1989a, b), Ishikawa (1976, 1988), Feigenbaum (1961, 1983) and Garvin (1983, 1984), have contributed to comprehensive quality management practices. Based on their research, industries have arrived at various tools and methods of

L. Wankhade and B. Dabade, *Quality Uncertainty and Perception*,
Contributions to Management Science, DOI 10.1007/978-3-7908-2195-6_1,

quality management. The methods used (Dahalgaard et al. 1998) during product development vary from the simple cause-and-effect diagram to the complex method of quality function deployment (QFD). For quality motivation and direct employee participation, team approaches to expert-oriented methods are practiced. Quality circles, control charts, experimental designs, etc., are adopted for quality management after the product is developed. Similarly, quality success of any industry or company can comprehensively be studied on the basis of critical evaluation of the major factors (Badri and Davis 1995).

Recently, many tools have been developed to boost up the quality endeavors. All the developments through the decades are encompassed by the newly coined term "Total Quality Management" (TQM). Now, quality is considered on a wider horizon. Customer satisfaction, customer demands, vendor proposals and similar exogenous factors are considered at the conceptual part of product design. Essentially, quality has become a part of concurrent product development.

Unlike product quality, service quality is an intangible entity. The concept of objective quality that fits into product quality has limitations to knowing the service quality, which is an abstract and elusive construct because of three features unique to services: intangibility, heterogeneity, and inseparability of production and consumption (Parasuraman et al. 1985). Here lies the origin of another term, "perceived quality." The mechanistic way of looking at the feature of a thing, event or object refers to objective quality and the humanistic way to perceived quality. Attempts to measure service quality have been made in the 1980s. Research work is still going on to formulate the metric for the measurement of service quality.

1.1 Recent Metrics for Business Performance

A few attempts to develop a metric that will be used to measure a company's performance at market end and, conversely, will be able to link the customer side to the company's growth have been made earlier. The earlier metrics are listed in the Table 1.1. These metrics are developed in different disciplines.

QFD is a development on the earlier process of offline quality control. In a product development phase, QFD embeds the customers' requirements of the product by using the matrix of house of quality. In the 1980s, the function of marketing evolved a technique of customer relationship management (CRM) which makes use of software to interact with the customers to elicit their responses. A continuous process of interaction may lead to better business strategies. Similarly, business performance can be evaluated by a balanced scorecard method.

The SERVQUAL scale was popular in the 1980s. In the following period, for want of a competitive advantage in businesses, the focus shifted to customer satisfaction. Recently, a new flurry of work has been noticed in management and business studies on the perceived value of a company. Customer satisfaction has been identified as a cause for the customer loyalty and company's performance (Andersson 1998). Attempts have been made to measure the level of customer

Table 1.1 Recent metrics for business performance

Metric	Meaning	Researcher and year	Discipline
QFD	A method to transform customer demands into product design	Yoji Akao (1966)	Quality management
CRM	Method to directly interact with customers to evolve customer based business strategies	Evolved in 1980s	Marketing science
BALANCED SCORECARD	Is a strategic performance management tool	Art Schneiderman (1987), Kaplan and Norton (1992)	Operations management
SERVQUAL	An instrument to measure service quality in terms of perceived quality.	Parasuraman et al. (1988)	Marketing science
CVA	A customer-survey methodology for business growth	Bradley Gale (1994)	Business management
NPS	Using customers' word-of-mouth for business growth	Fred Reichheld (2003)	Business management
CRM–TQM	Using CRM to improve the organization's TQM culture	Curry and Kkolou (2004)	TQM

satisfaction (Keiningham and Vavra 2001). Bradley Gale introduced Customer Value Analysis (CVA) as a similar measure. Fred Reichheld developed the Net Promoter Score (NPS) in 2003. The NPS has been received with great enthusiasm by managers because of its simplicity and the way the metric links the customers' word-of-mouth power with the company's growth.

But Keiningham et al. (2008) critically examined the NPS and noted "There are myriad dimensions affecting customer loyalty and how it impacts customer behavior and profitability. When simple solutions are sought to such a complex problem, the answers are often technically correct in a narrow sense – but substantially wrong. Net promoter, like any measure of customer intentions, is inherently unreliable."

From QFD to NPS to CRM–TQM are the milestones in the development of metrics. But the process of metrics formulation has been mostly lopsided. The metrics were confined to any single discipline or were based on a few selected issues. Complete or near-complete perspective to the metric has not been realized any time. Similarly, TQM lacks the strategies' dimension that should have been useful to deliver business performance.

1.2 Quality Pursued in Various Disciplines

The concept of quality is equally applicable to products and services. And the research on the aspects of quality has been pursued differently in different disciplines. Various disciplines – economics, manufacturing, management and marketing – have their own parlance, many times resulting in conflicts (Garvin 1984). Currently, the studies on quality aspects are pursued separately in economics, management and marketing, as shown in Fig. 1.1.

Specifically, the work possesses the tone of the discipline in which it has been pursued. Moreover, the differences in approaches have grown further to a large gap. But every study has a significant contribution. These approaches, though isolated, are important for the holistic understanding of quality. Now, it is a ripe time for these approaches be assimilated so as to evolve a sound approach to quality management.

The first attempt of this nature, where studies on the aspects of quality across different disciplines are considered, was made by Garvin in 1984. He beautifully assimilated all these approaches and arrived at the eight dimensions of quality. Perhaps, for the first time quality has been defined in a realistic sense. It offered a multidisciplinary approach to defining the term "quality."

Again, research is traversing through paths of isolation. And, once again the need has been established that quality should be looked into from all these perspectives. Figure 1.2 presents a streamlined approach in which the best quality management practices would be possible by using all the existing studies. This perspective may provide the required vibrancy which can stand up to the competitive spirit that has been emerging across the world.

Before arriving at a multidisciplinary approach, let us take a glance at the studies on quality that have been done in various disciplines.

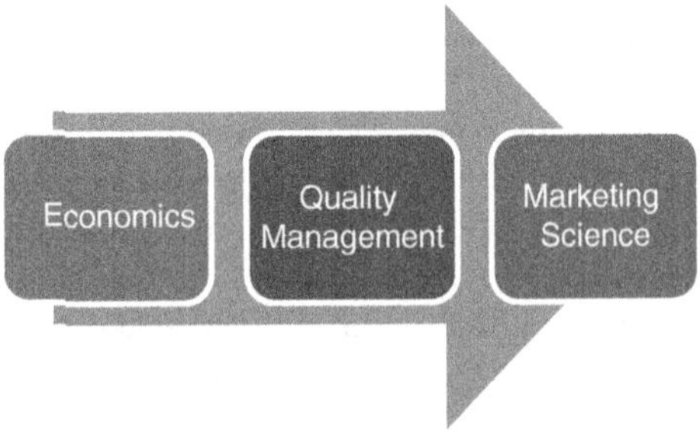

Fig. 1.1 Quality pursued in isolation

Fig. 1.2 Streamlined approach for studies on quality

1.2.1 Total Quality Management

TQM is an encompassing term where quality has been incorporated as a part of the management philosophy. TQM is a philosophy and its implementation varies from industry to industry. Study of TQM, its conceptual understanding and its implementation are more appropriate if carried out across continents or nations.

Brown (1995) had coined the term "nation equity" for the perceived quality of a product from an individual nation's point of view. He has compared nationwide quality on the basis of a poll and attempted to see the perception of UK product quality across the world. Brown has drawn the important conclusion that quality perception will be increasingly the determining factor as consumers become capable of purchasing almost any product from many countries and manufactured by companies facing similar pricing pressures.

National culture is being identified as an important determinant of quality. Noronha (2003) in his study revealed that the cultural traits in China, Hong Kong and Taiwan could very well be intertwined to some of the quality dimensions. Prasad and Tata (2003) have studied the role of socio-cultural, political-legal, economic and educational dimensions in quality management. They have developed a relationship between these international conditions and quality dimensions which are helpful in understanding the differences in quality practices across countries.

Jabnoun (2000) identified the need of restructuring for TQM. Ismail and Maling (2002) aptly observed "an analysis of the TQM studies conducted in various countries revealed that there is a lack of information about the nature and stage of implementation in other regions of the world including South America, Africa and the Middle East." Many of the critical factors are identified but without sufficient elaboration. Ismail and Maling have stated that factors such as strategic planning, product and service design, communication, social responsibility, employee appraisal, rewards and recognition need to be given due attention. They have also commented that there is absence of a universally accepted TQM model so far and more studies are required to understand the quality paradigm.

Motwani (2001) elaborated the critical factors and performance measures of TQM. He carefully studied various authors (Ahire et al. 1996; Black and Porter 1996; Flynn et al. 1994; Powell 1995; Saraph et al. 1989; Zeitz et al. 1997) and observed "some authors focus on the technical and programmatic properties of TQM, while others look at the general management philosophy." Critical factors are something like constructs which can build TQM like constructing a house. Of the seven identified critical factors through the studies of various authors, Motwani identified *top management commitment* as the base or foundation of TQM, along with four other factors as four pillars and the remaining two as the final elements.

McAdam and Henderson (2004) investigated the scope and depth of the influencing or driving factors that will shape the future of TQM. They have classified TQM drivers as internal and external. Curry and Kkolou (2004) demonstrated the customer focus that TQM should have, by considering the CRM–TQM paradigm. Some standardized models – the Malcolm Baldrige National Quality Award model

Table 1.2 Tools and techniques for quality improvement

Tools	Techniques
Acceptance sampling	Analysis of variance
Activity network diagram	Benchmarking
Affinity diagram	Capability measures and ratios
Brainstorming	Cross-functional team
Causal loop diagram	Design of experiments
Cause and effect diagram	Exploratory data analysis
Check sheet (tally sheet)	Failure mode and effects analysis
Control chart	Fault tree analysis
Flowchart	Matrix data analysis
Force field analysis	Modeling and simulations
Graphs	Optimization techniques
Histogram	Poka yoke
Inspection	Process discovery
Interrelationship diagraph	Quality costing
Language processing diagram	Quality function deployment
Matrix diagram	Queuing theory
Pareto diagram	Regression analysis
Prioritization matrix	Rejection analysis
Process decision program chart	Response surface method
Questionnaire	Root cause analysis
Relations diagram	Statistical quality control
Reliability block diagram	
Run chart or record	
Scatter diagram	
Split-plots	
Tree diagram	

in the United States, the European Foundation for Quality Management (EFQM) model in Europe and the Deming Application Prize model in Japan – act as a guide for better quality practices.

Further work on TQM structuring was done by Tar´ (2005) who suggested that not only critical factors but a set of components consisting of critical factors, tools, techniques and practices should be used for successful implementation of TQM. Other TQM tools such as flowcharts, cause-and-effect diagrams, Pareto charts, histograms, run charts and graphs, X-bar and R-control charts and scatter diagrams are still useful. Dabade and Ray (1996) and Dale and McQuater (1998) have identified the tools and techniques most widely used by firms. The tools and techniques used for quality improvement that play important roles alongside TQM critical factors are shown in the Table 1.2.

1.2.2 Economics

The studies in economics are either macroeconomic or microeconomic, which is decided by the nature of the problem being handled. But it is the transition from the old to the new economics that has changed the whole perspective that is offered for the economic study of any issue. "The old economics was based on a variety of prejudices" Akerlof (2001). It considered the model of equilibrium along with complete information for market transactions.

Market equilibrium is a market condition with equality of the quantity supplied and the quantity demanded. It refers to the situation where the supply of an item is exactly equal to its demand. Equilibrium presumes a market without shortage or surplus and hence no possibility of change in the price of the product. Similarly, for product information, market equilibrium assumes equality of information with sellers and buyers. This notion of complete information, thus, becomes the basis for market transaction in market equilibrium.

Quality embodied in transacted goods or services in any market has been studied and the economics was developed for its evaluation. It has been the change to modern economics in which the studies on quality have gained the momentum. Solow changed the pattern of the study. He developed the type of framework in which goods varied by quality. "But there was no analysis as to how different qualities of goods would affect markets" Akerlof (2001).

1.2.2.1 Information Economics

Later on, in modern economics, the concept of equilibrium and complete information in a market place was refuted. The old theories have seen a tremendous change. New theories based on information imbalance or information asymmetry have been developed. In a market transaction, one party may have better information than the other. This causes information asymmetry between the buyer and the seller. Either the buyer or the seller may have more information and hence will be in a beneficial position in the transaction. Many informationally asymmetric conditions are observed in a large number of real-life institutions. Specifically, business world suffers from this malaise.

Quality uncertainty due to information asymmetry is another quality dimension from the modern economics of nonequilibrium. While recognizing the importance of the phenomenon of quality uncertainty that has been discovered in information economics, Ali and Seshadri (1993) commented that analysis allowing for the distinction between "objective" and "perceived" quality has not received much attention among economists. Moreover, the concept of information has not been applied to perceived quality. Shapiro (1982) suggests that the way choice of product quality related to the information-gathering activities of individual customers, or the information flow in the market place, remains an under-researched issue.

Akerlof (1970) discovered the phenomenon of quality uncertainty that has revolutionized the research on quality in the field of economics. A vast literature demonstrating the uncertainty of quality has been available since then. Resale transactions have found more space in the literature on information economics while demonstrating the phenomenon of quality uncertainty, though it can be applied to all types of goods and all types of transactions. Only the extent of the effect of information asymmetry on quality uncertainty can be different – minimal, moderate or maximum. The theories of information asymmetry are equally applicable to the manufactured products which are immediately received in a market. The domain of primary sales is significant for the dimension of quality uncertainty and is more appealing in manufacturing parlance than in the resale market.

1.2.3 Marketing

Customer is the focus of the marketing science approach to quality. Importance is given to the way customers receive products or services. Perceived quality is an attraction in the marketing science approach to quality. Thus, the humanistic approach overshadows the mechanistic one. Objective quality passes through various marketing parameters to become perceived quality. The mechanism of quality perception is observed in the literature.

The factors that are responsible for such a perception are also well studied. Many such factor-specific studies have appeared in the literature. Researchers have attempted to understand the effect of advertising on consumers. Similarly, the other factors like market share, reputation, price, word-of-mouth, brand name and financial information are considered for the evaluation of quality. This has helped understanding the possible factors that cause quality perception.

But, most of the marketing science literature relates to service quality. The metric for the measurement of service quality has been developed as SERVQUAL in the 1980s. Refinements to SERVQUAL and further research work toward the metric formulation are still in progress. Only very little work has been carried out on product quality perception. This is supported by the fact that goods quality can be measured objectively by such indicators as durability and number of defects (Crosby 1979; Garvin 1983).

The emergence of stiff competition in recent times has emphasized the need of learning the mechanism of product quality perception. Some attempts have also been made to see the reflection of product quality through customers' eyes. The data are collected on the basis of product categories and statistical observations. Marketing science approach presumes consumer expectations of product quality. Perceived quality is then computed on the basis of such findings.

An approach similar to SERVQUAL has been practiced for measuring product quality (Ali and Seshadri 1993). Reference to a standard is an important parameter in such a perception. Quality is perceived on the basis of some reference rather than the objective quality itself. The terms like "accurate perception" and "imprecise perceptions" are used in the reference-based process of perception. Thus perceived quality is largely based on the reference standards that customers hold while evaluating quality.

Thus, it has been observed that studies on quality are again traversing through paths of isolation. Figure 1.3 shows the effect of quality uncertainty due to information asymmetry (the lemon problem) felt in the major areas – economics,

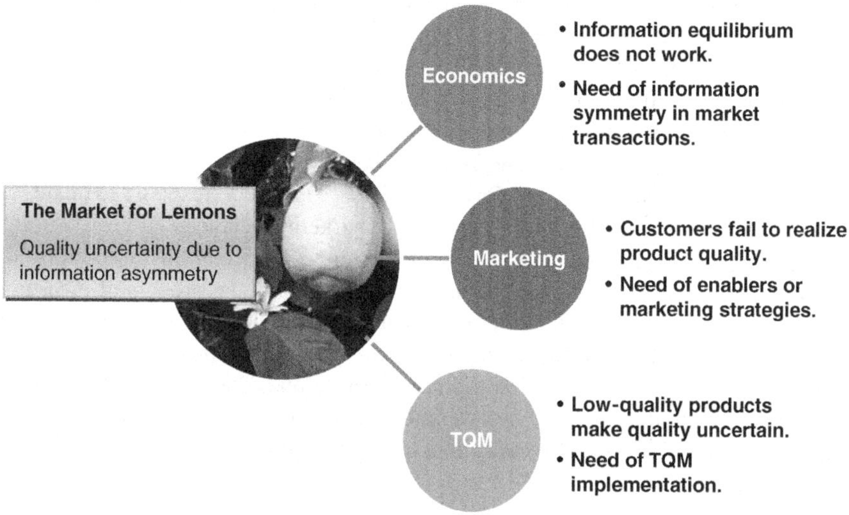

Fig. 1.3 Effect of quality uncertainty on major areas

marketing and TQM – and further corrective actions that need to be taken in these areas.

Information economics deals with information nonequilibrium in the market that has been caused by information asymmetry. It needs to be corrected by making information more symmetric. Similarly, customers are unable to know the product quality in an informationally asymmetric market. Marketing should play a role to launch marketing enablers so that customers are able to understand the true quality. The origin of quality uncertainty lies in the manufacture of low-quality products, and this can be corrected at source by promoting the implementation of TQM culture in industries. Thus, the reflection of quality uncertainty is clearly observed in economics, marketing and TQM.

1.3 Quality Commonality and Multidisciplinary Approach

A careful review of the current research in these disciplines reveals some common thread that seems interwoven. The research in TQM has been growing outwardly. TQM–CRM nexus to socio-economic-cultural factors are indicative of the customer orientation to the TQM research. Specifically, TQM constructs have been designed in the backdrop of economy, culture and consumers. The lacking universality of the TQM model can be ascribed to the change of culture from nation to nation.

Earlier, research in quality management had its focus on improvement of objective quality where the mechanistic aspect of quality predominated. With TQM coming to the fore, the humanistic aspect is being accommodated. Gradually, quality improvement is becoming an inherent part of TQM, and the focus of study has been shifting to the humanistic aspect.

As TQM traverses from a mechanistic zone inside the company to a humanistic zone at the market place, it is entering the marketing science domain. Many aspects of quality study are becoming common to both TQM and marketing science. Now market is a common place where TQM as well as marketing science studies on quality intersect. Specifically, perceived quality as customers' response to quality is equally identified in quality management and marketing science.

TQM developments have taken care of customers' expectations in the form of QFD or some TQM constructs. The duplication of this in the process of measurement of perceived quality in marketing science can be ascribed to the isolation of marketing from quality management, though intermittent efforts have been made to link quality management function to the marketing one. The attempts to measure acceptance of product quality by customers should then be based on the efforts made for marketing and its realization. And the reference standard can be the objective quality itself.

At this juncture, the field of economics has experienced many changes and similarly the approach to studying quality has changed. The emergence of economics of information in recent times has added a new dimension to the study. But, the presence or absence of product information at the market place, as a basis for quality evaluation, is seen largely absent in quality management and marketing science.

Quality uncertainty is a recent dimension in information economics. Information asymmetry makes quality uncertain. But the literature on economics is silent about the quality perception and the required role of information. On the contrary, marketing science has much to offer on quality perception but is devoid of information content. In marketing science, the quality uncertainty part is silent. Whereas quality management has been showing market orientation, its philosophy of TQM is aloof from information asymmetry and quality uncertainty.

It has been suggested that TQM-based approaches are failing to deliver the anticipated business performance improvements in many companies (Greising 1994; Jacob 1993). It might be due the fact that quality management techniques were excessively focused on internal processes. Empirical evidence suggests that business performance improvements may result from quality strategies (Aaker and Jacobson 1994) and quality strategies cannot be developed by improving the internal processes only. It should be conjoined to marketing so as to make it a strategic tool. Specifically, marketing inputs play a crucial role in knowing the customers' expectations and accordingly quality management or TQM can be reoriented. The dimension of quality perception recognizes the translation of objective quality to the market worth and also senses the customers' requirements from the market end. Thus quality perception as a dimension of quality management embeds strategies that can be used to improve business performance.

Morgan and Piercy (1998) have studied the interactions between marketing and quality in the backdrop of strategic business units. Customer-focused quality is a "quality revolution" in the business world. Quality uncertainty and perceived quality are the customer-focused dimensions that are identified as the most common and recent, noticed in economics, marketing science and quality management. These dimensions that are present in various disciplines should have a collectively representative metric. Thus the metric should be able to measure the acceptance level of the quality at the market place. And the metric should also be able to improve the level of such acceptance.

Emerging quality dimensions in all these disciplines tend toward directions that seem similar but are different in content. The approaches followed in these disciplines are incomplete, as these were pursued mostly in isolation of each other. It seems appropriate to bridge quality management with marketing science by using the connection of information economics. Assimilation of these three disciplines will certainly evolve a robust theory that could accommodate all viewpoints. And now there is a re-emerging need of a multidisciplinary approach to quality.

1.4 Management of Quality Uncertainty and Perception

The earlier metrics tried to compute numbers for improving business performance. The metrics definitely had a substance of truth to make the change happen. But mostly the metrics were based on partial understanding and hence these led to only partial success. Hence, attempts are being made to devise approaches rather than a complete solution. The multidisciplinary approach uses information basis from the related disciplines and tries to provide near completeness to the issue of perceived quality.

As shown in the Fig. 1.4, the approach uses the recent developments – quality uncertainty from information economics, perceived quality from marketing science and TQM from management. These developments are used for the management of quality uncertainty and perception – to minimize quality uncertainty or to maximize quality perception – which ultimately makes business grow.

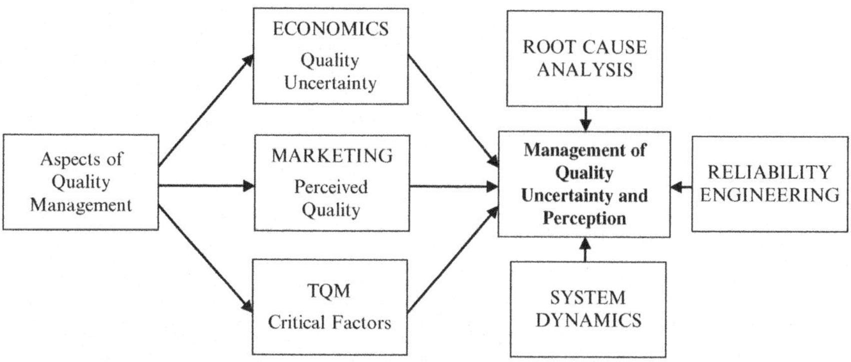

Fig. 1.4 Management of quality uncertainty and perception

As shown in the Fig. 1.4, the concepts of quality uncertainty, perceived quality and TQM critical factors (TQM constructs) are presented separately, as studied in information economics, marketing science and TQM respectively, before these are merged together. Earlier studies on quality uncertainty and perception were pursued without assigning any relationship to each other. New definitions are coined and relationships are established then by assimilating quality uncertainty, quality perception and TQM. For the process of assimilation, one more discipline of study has been used – probability and reliability engineering with success and failure analysis – as a basis for renewed theory building. Root cause analysis presents an approximate approach where an analyst can minimize quality uncertainty or maximize quality perception by using the root cause method from management science. The most refined approach is the one in which modeling and simulations by using system dynamics can achieve the highest level of performance improvement. The system-dynamics-based approach is particularly important as it provides many strategic options.

Chapter 2
Quality Uncertainty Due to Information Asymmetry

2.1 Prelude to Quality Uncertainty

After the Nobel Prize was awarded to Akerlof, Spence, and Stiglitz for their contributions related to market with asymmetric information, the term "information asymmetry" has became known in every nook and corner of the world. The theory of markets with asymmetric information rests firmly on the work of three researchers: George Akerlof (University of California), Michael Spence (Stanford University), and Joseph Stiglitz (Columbia University). The Royal Swedish Academy of Sciences (2001) in its notification for the Nobel Prize said "Akerlof, Spence, and Stiglitz's analyses form the core of modern information economics. Their work transformed the way economists think about the functioning of markets. The analytical methods they have suggested have been applied to explain many social and economic institutions, especially different types of contracts."

Economics of information covers a wide range of issues across the market spectrum. High interest rates in local lending markets in the Third World, firms paying dividends even if heavily taxed, mix of insurance premiums, landowners and sharecroppers, etc., are the conditions where the information is present in disguise (Royal Swedish Academy of Sciences, October 10, 2001). Information economics deals with information asymmetry and its associated repercussions on the market along with cost–benefit analysis and plays a pivotal role in understanding the performance of businesses.

Akerlof as a young man of 27 spent a year in India. In his own words, "after being to India I understood that economic systems don't necessarily work as they do in standard economics, where markets always clear. The caste system somehow provided me with an alternative model for how economic systems might work." He elaborated this observed phenomenon and coined a term "information asymmetry." It has been noticed that information asymmetry prevails everywhere, and the difference lies in the extent of its incidence. Akerlof (1970) explained this phenomenon in his classic and seminal piece "The market for lemons: quality uncertainty

L. Wankhade and B. Dabade, *Quality Uncertainty and Perception*,
Contributions to Management Science, DOI 10.1007/978-3-7908-2195-6_2,

and market mechanism." His paper was rejected thrice by reputed international journals before the *Quarterly Journal of Economics* accepted it. Akerlof (2003) narrated this and related experiences in "Writing the Market for 'Lemons': A Personal and Interpretive Essay." In this essay, he elaborated the changing scenario of economics.

Akerlof wrote: "The old economics was based on a variety of prejudices. The first of these was the primacy of the general equilibrium competitive model with complete information. That did not mean that all economists believed that this model was a particularly good description of markets and how they operated, but it did give them a benchmark from which to measure the consequences of departures from its strict assumptions. In this way economists used the competitive model as the major road map of their world." An important part of Akerlof's research is that market does not always behave in equilibrium and may work in imperfect competition due to the asymmetry of information.

2.2 Information Asymmetry and Quality Uncertainty

Stiglitz (2001) defined information asymmetry as "fact that different people know different things: workers know more about their ability than does the firm; the person buying insurance knows more about his health, whether he smokes and drinks immoderately, than the insurance firm; the owner of a car knows more about the car than potential buyers; the owner of a firm knows more about the firm than a potential investor; the borrower knows more about his risk and risk taking than the lender." So, it is a condition in which at least some information is known to some but not to all parties involved.

Asymmetry of information occurs when one party in a transaction has more or better information than the other. Examples are used car sale, stock broking, sharecropping, real estate business, and life insurance transactions. Usually, the seller has more information about his or her product than the buyer although reverse may also be true. This situation was first explained by Kenneth Arrow in 1963.

Akerlof has illustrated the case of information asymmetry by using an example of used cars in the resale market. Defective cars are called as "lemons." Here, the word "lemon" can be used for any low-quality product that drives out a high-quality product from a market. *When the buyer is less informed than the seller about the quality of the product, he/she would refuse to pay the price for good quality since the product bought might turn out to be bad (i.e., the uncertainty about product quality due to information asymmetry). Then the seller will also have no incentive to make good-quality products available. In other words, the bad-quality product will push the good-quality product out of the market (quality uncertainty syndrome).* Akerlof generalized this in a rigorous way so that a variety of situations involving "thin" markets could be explained. (A market with a small number of buyers and sellers with few transactions is called a thin market.)

2.3 Examples of Information Asymmetry and Quality Uncertainty

The phenomenon of quality uncertainty due to information asymmetry has been predominant in the resale market where high-quality products are called as "plums" and low-quality products "lemons." A car owner may sell his car, but his real intentions and condition of the car may not be made known to the potential buyer, which leads to information asymmetry. Numerous examples are seen where sellers or buyers have more information. A few examples are given in the Table 2.1.

2.3.1 Car Seller

A seller of a car has true information regarding its quality. Consider it is a lemon. The seller has no interest in revealing the information that his car is a lemon. If he does so, he will improve the information of the buyer and will lead to information symmetry between him and a buyer, regarding the quality of his car. Then he receives a lower offer for his car. If he decides to keep the information for himself, he will create information asymmetry between him and a buyer. In this situation, the buyer will think of the possible qualities of the offered cars in the market. The buyer will assume an average price (of plums and lemons available for sell) for the lemon car. Thus the owner of a lemon will sell his car at higher price in asymmetric condition.

Now, consider that car is a plum. The owner of a plum has interest in revealing the information that his car is a good-quality car. If he convinces a buyer that he has a plum, there will be information symmetry and he will get an appropriate price for his car. But, it is a difficult task to improve the information of a buyer as the market is drowned by the noise of owners of lemons. And thus asymmetry prevails again.

Table 2.1 Some examples of information asymmetry

Sellers with better information		Buyers with better information	
Example	Meaning	Example	Meaning
Used car sellers	A car seller knows the "lemon" and "plum" cars	Medical insurance	Insurance buyer knows the exact health status
Stockbrokers	Stockbrokers may know the true value of a company	Employee or worker	Job seeker knows his/her capabilities and motivational level for the required job
Real estate broker or agent	They know the true value of the saleable properties	Sales of old art pieces without assessment	The buyers may have better judgments about the value of the art
Online shopping	Sellers only know the quality of the items	Home owners	Home owners know their property expansion and liable taxation than authorities

The offered price for a plum is the average price of plums and lemons. The seller has no interest to sell his car at a lower price and hence prefers to exit the market.

2.3.2 Insurance Policy

Insurance firms calculate policy premiums by considering low- and high-risk clients. Insurance policy charges an average premium where the exact health status of a client is not known. Thus an average premium reflects the cost of low- and high-risk clients. This is a case of information asymmetry. A low-risk customer finds the price heavier and declines to purchase the policy. On the contrary, a high-risk client does not reveal the real health status and purchases a policy, making it a lemon situation. Therefore, it is required that information related to the health status of a customer should be ascertained by using an appropriate mechanism to calculate the premium. This creates information symmetry where policy premiums are real.

2.3.3 Information Technology (IT) Firms and Investors

In a new sector, such as IT, most firms may appear identical in the eyes of the uninformed investor, while some insiders may have better information about the future profitability of such firms. Firms with less than average profitability will therefore be overvalued by the stock market where, of course, uninitiated investors also trade. Such firms will therefore prefer to finance new projects by issuing new shares. Firms with higher than average profitability, on the other hand, will be undervalued and find it costly to expand by share issue.

2.3.4 Online Shopping

Some online shopping places are the markets for lemons as they are not interactive. The information of articles kept for sale can be intentionally scanty or superficial or fabricated. A buyer orders the item and discovers that it is a lemon when he receives it. Many fraudulent selling practices are noticed on these sites. For example, a seller opens an account on a typical online shopping site. He has low-capacity MP3 players (say 1 GB). He hacks the software so that when plugged into a computer, it declares the players as high capacity players (2–3 GB). He sells it at the price of 2–3 GB, making it a lemon's market.

The phenomenon of quality uncertainty is also true in primary sales. The difference lies in the level of information asymmetry and corresponding quality uncertainty. The level can be high in the resale market and can range from high to low in primary sales. Hence, the level of quality uncertainty depends on information asymmetry.

Information asymmetry is also dependent on the type of product. The information asymmetry then becomes product information asymmetry for primary sales. Thus, quality uncertainty may be low, medium, or high. Products like suits, leisure wear, air coolers, books, pencils, and pizzas contain low information asymmetry, whereas watches, TVs, cameras, and cell phones have medium information asymmetry. Soaps, detergents, shampoos, paints, and toothpastes possess high information asymmetry.

Quality uncertainty can be redefined for product sales:

- Quality lost due to the asymmetry phenomenon, in transmission from the source to the market
- Probability by which customers fail to realize the product quality
- Probability by which the product quality does not get translated to a market value

2.4 Implications of Asymmetric Information

The foremost impact of asymmetric information is adverse selection. Under asymmetric information, the "low-quality" firms tend to grow more rapidly than "high-quality" firms, implying that the market will be gradually dominated by "lemons." The characteristics of the lemon market are as follows:

- The seller knows the true quality, but the buyer does not
- The buyer is willing to pay based on the average or expected quality
- Low-quality products ("lemons") may drive out high-quality products

Akerlof (1970) has suggested some mechanisms such as guarantee, warranty, reputation, and licensing practices to counteract the lemon problem. These remedies can be simplified to the following:

- Sellers convey information to buyers through informative advertising, building reputation, or offering credible guarantees or warranties.
- Industry groups, governments, or consumer advocacy groups can establish minimum standards and/or provide certification.
- Trusted agents can provide detailed quality information.

Spence suggested the strategy of signaling to overcome adverse selection. It is an observable action taken by a seller to convince the buyer of the product quality. But the opportunity cost must have a sufficient difference for its success. The buyer adjusts the purchasing behavior as per the signal. High pricing of a product is a popular example of signaling the high quality of a product.

Another outcome of a market with asymmetric information is "moral hazard." Moral hazard is the consequence of asymmetric information after the transaction occurs. For example, warranty is normally offered as a signal of the high quality of a product. However, reckless or wanton use of the product may defeat the purpose of offering warranty and result in moral hazard.

In the market with asymmetric information, normally, it is the seller who knows better than the buyer and has an advantage. The advantage may be translated to the unscrupulous behavior of the seller where he may engage the transaction or sell the product/service of lower quality at higher prices pretending that the product/service is of higher quality. Akerlof (1970) has also considered the cost of dishonesty in "lemons" market. He states: "Dishonesty in business is a serious problem in underdeveloped countries. The cost of dishonesty therefore lies not only in the amount by which the purchaser is cheated; the cost also must include the loss incurred from driving legitimate business out of existence."

Thus, when the customer is equally informed he/she will buy high-quality products (of course, depending on his/her economic condition). However, the market consists of informed as well as uninformed customers. Uninformed customers may buy low-quality or high-quality products. But, in lemon circumstances high-quality product manufacturers may signal the product by high pricing. Then the chances are that uninformed customers will be easy prey to lemons.

2.5 Product Manufacturing and Quality Uncertainty

Acceptance of a quality product (manufactured or service) is influenced by the informational rubric in the market. Low-quality products (lemons) penetrate the market at the cost of high-quality products. Informationally asymmetric customers perceive the product price as an average of low and high, which subsumes lower to the cost of manufacturing of high-quality product, resulting in quality uncertainty.

The mechanism of quality uncertainty due to information asymmetry for a product can be explained by modeling a simple example. Assume that products are sold in indivisible units and are available in two qualities, high and low, with market share of λ and $(1 - \lambda)$, respectively. We presume that each customer will purchase one unit and there is only one manufacturer for each quality. All the buyers have same valuation of these qualities, low and high. One product of low quality is priced at P_L rupees and one product of high quality is priced P_H rupees, with $P_H > P_L$. Also, each manufacturer knows the quality of his product. Production cost of low-quality product is C_L ($< P_L$) rupees and high-quality product is C_H ($< P_H$) rupees. Potential customers, either by forecasting or past sales, establish the market size of any specific product. The part of this market size based on the sales volume per unit time makes the "Market Share" of any industry. Profit share connotes merely the profit margin or amount added over the cost of production for making the price of the product.

2.5.1 Market with Complete Symmetry

Here, customers distinguish between the low and high quality. As shown schematically in Fig. 2.1, product price for both manufacturers is higher than the cost of

Relationship: Manufacturer		Relationship: Customer	Market Share	Profit Share
C_H	<	P_H	0 λ 1	0 C_H P_H
C_L	<	P_L	0 (1-λ) 1	0 C_L P_L

Fig. 2.1 Market and profit share for complete symmetry

Relationship: Manufacturer		Relationship: Customer	Market Share	Profit Share
C_H	>	P_A	0 λ=0 1	0 P_A (–) C_H
C_L	<	P_A	0 (1-λ)=1 1	0 C_L P_A

Fig. 2.2 Market and profit share for complete asymmetry

production. Products with acceptability worth their prices are able to create their own market share. Considering separate market for low and high quality, price between C_L and P_L will benefit both customer as well as manufacturer. Given that the market is fully symmetric, the extent of the market of the high-quality product depends on the economic lot of the people. Higher income with better purchasing power parity drives wider acceptance of high-quality products and may drive out the low-quality products. Thus, λ is a function of the socioeconomic development of the nation.

2.5.2 *Market with Complete Asymmetry*

In developing nations, when asymmetry prevails, customers are unable to perceive the quality. Customer valuation of the product becomes the average of the two quality levels and they assume that market price should not exceed $P_A = \lambda P_H + (1 - \lambda) P_L$. For high-quality products, populace's perceived price P_A is less than the cost of production C_H, making a loss in the product sale and hence their market share dwindles to zero or negative. On the contrary, low-quality product manufacturer hails the advantage in increased market share, without any value addition and at the cost of high-quality product manufacturers. As shown explicitly in Fig. 2.2, the former have grabbed the total market with expanded profit share than earlier (compared to Fig. 2.1) driving the latter out of the market.

2.5.3 *Market in Developing Nations*

Both distributions (Figs. 2.1 and 2.2) are ideal and nonexistent in the real world. Practically, high-quality products are not driven out of the market, as is never the

Relationship		Market Share	Profit Share
Manufacturer	Customer		
C_H < C_H >	P_H P_A	(–) --|-- (+) Fluidity Zone ....|---...................... – 0 λ 1	...|...............|------|----... 0 P_A(–) C_H P_H
C_L < C_L <	P_L P_A	Fluidity Zone (–)--|-- (+) ....|--------------------|.... – 0 (1-λ) 1	...|...------------|............ 0 C_L P_L P_A

Fig. 2.3 Market and profit share in developing nations

fully asymmetric one. But given the fact that the phenomenon of asymmetry is significant in developing countries, the impact on product quality (i.e., uncertainty) is sufficiently high. Figure 2.3 shows the symmetric and asymmetric combinatorial market distribution on approximate assumptions.

The market in developing nations consist of both low-quality and high-quality products. The prevalence of low-quality products is significant there. This can be attributed to the presence of the large extent of information asymmetry and the small extent of information symmetry. In developing nations, the larger part of the customer population is informationally asymmetric. They assume product price as an average of high- and low-quality products. Then, the expected product price becomes the average price. In this case, a larger population will believe in expected price which is lower than the high-quality product cost or price. Hence, larger asymmetric customers dwindle to the low-quality product.

Still, sizable population of customers is informationally symmetric and knows about the higher price of the high-quality product. They believe in the high-quality price and hence buy the high-quality product. This is shown graphically in the form of the market share. The market share of the low-quality product is proportional to the larger size of the customers being informationally asymmetric. The market share of high-quality product is smaller, which is also proportional to the smaller extent of prevailing information symmetry. Similar reflections are observed on the profit share of low- and high-quality products in the backdrop of an average or a perceived price.

So, a low-quality product manufacturer gains from market shares, low as well as high. A schematic representation of the changes in market and profit share with respect to Figs. 2.1 and 2.2 is shown in Fig. 2.3. There is always a small segment of the population vulnerable to symmetry as well as asymmetry. A small segment of the population may become symmetric in due course of time, whereas some portion may become asymmetric during the same course of time. These dwindling customers are represented by the fluidity zone in Fig. 2.3.

2.6 Causes of Quality Uncertainty due to Information Asymmetry

In the first place, it is important to know why there might be information asymmetry. It refers to lack of information with a party in a transaction and a few questions are posed to elicit the causes of its occurrence.

Fig. 2.4 Informational position of a buyer

- What makes the lack of information?
- Is it that a buyer is not exposed to the necessary information?
- Is it because a buyer has not shown interest to acquire the required information?
- Is it because desirable information is not available?
- Is it due to the fact that a buyer has been socialized in such a way that he never cares for such information?
- Is it that a buyer does not have the time to gather the information?
- Is it that gathering of information needs money?

Each question has relevance to information asymmetry. The buyer's informational position in the backdrop of possible causes is also shown in Fig. 2.4. The causes are specific as well as general. Mostly, the desired information is not observable. And if the information is tangible, then it takes time for acquiring it. Thus, it is costly to obtain, store, and process the information.

Figure 2.5 shows the mechanics of quality uncertainty due to information asymmetry. Broadly, as shown in the Fig. 2.5, presence of low-quality products, prevailing socioeconomic conditions and lack of specific marketing efforts from the manufacturers of high-quality products result in information asymmetry at the market place. Information asymmetry in turn develops quality uncertainty to the high-quality products. It then goes into a vicious circle, making the phenomenon of quality uncertainty due to information asymmetry.

Information asymmetry has been observed to be predominant in developing nations. It subscribes to the fact that there are basic differences in having information asymmetry at different places. Information asymmetry can be broadly considered into two types – product information asymmetry and general information asymmetry. Specific causes can be attributed to product information asymmetry, whereas general causes are attributed to general information asymmetry. The buyer's position due to location or socioeconomic conditions determines his general information asymmetry. Thus, information asymmetry that leads to quality uncertainty has two subsets of causal factors:

Fig. 2.5 Mechanics of quality uncertainty

1. Socioeconomic and cultural conditions
2. Enablers or marketing strategies

2.6.1 Socioeconomic and Cultural Conditions

2.6.1.1 Education

Education is the most important factor in understanding information asymmetry and the associated quality uncertainty. Education prepares the mental faculty for logical development as well as for receiving, processing, and analyzing information. Education acts as a basic driver for information (imperfect to perfect) flow. The greater the educational attainment of the people, the greater is their drive toward information perfection. At the micro level, education can have ramifications in the field of specialization and in how the broader knowledge spectrum is being developed. Knowledge specialization explains information perfection or imperfection and the associated quality uncertainty.

2.6.1.2 Cultural Milieu

The cultural milieu of a society is next in importance. As a part of the process of socialization, culture plays a pivotal role in shaping the personality type. No culture is superior or inferior; but cultural differences among human societies are remarkable. Cultural studies are undertaken by anthropological scholarship; however, our pragmatic aim is to recognize cultural differences and to study their impact on quality uncertainty.

2.6.1.3 National Economy

Economic causes contribute more toward the asymmetry phenomenon and ultimately lead to low-quality product acceptance. Some of the asymmetries arise naturally out of the economic processes (Stiglitz 2001). Economy is always classified by epithets such as agricultural, service, industrial, or mixed. This classification connotes the profession or economic activities of the majority of the people. Also, economy implies the populace status, livelihood, and their inclinations. The relationship between economy type and information asymmetry cannot be directly established, but, underneath, economy is strongly responsible for quality uncertainty.

2.6.1.4 Global/National Enactments

Businesses are governed by the laws of the nation in which they are conducted. Normally, appropriate business certifications are required where the codes of conduct are mentioned. Curbs are imposed on the fraudulent business practices by national or international enactments. In many places, laws such as liability, antifraud and disclosure laws are practiced. For example, disclosure law is used where sellers have to disclose the all related information. These laws help solve the problems created by asymmetric information.

2.6.2 Enablers or Marketing Strategies

2.6.2.1 Efforts for Quality

Continuous efforts from product development to manufacturing are essential for better quality of the product. Methods used (Dahalgaard et al. 1998) during product development vary from a simple cause-and-effect diagram to the complex method of QFD. For quality motivation and direct employee participation, team approaches to expert-oriented methods are practiced. Quality circles, control charts, experimental designs, and so on are adopted for quality management after the product is developed.

Now a holistic philosophy has been developed called TQM. Quality success of any industry or company can be comprehensively studied on the basis of a critical evaluation of identified critical factors (Badri and Davis 1995). These are the TQM constructs. These constructs are implemented through various performance measures. Every TQM construct has a few performance measures. Implementation of these measures determines the TQM status of any industry. And the TQM status ultimately decides the product quality.

2.6.2.2 Supply Chain Management

A supply chain is a set of structures and processes that an organization uses to deliver an output to the customer (Sterman 2000). The supply chain ambit covers a variety of issues such as physical product as well as resources including skilled labor, services, and product design. Product quality depends, in the first place, on raw material supplied, parts or subproducts obtained from subcontractors, control software and the workforce recruited. The smooth behavior of the supply chain directly or indirectly helps maintain product quality. Supply chain is a back-end operator.

2.6.2.3 Company Reputation

The track record of any product in the market builds the reputation of the manufacturer. Consistently offering better quality products in any particular category definitely pays back. A company can garner reputation as high as when customers start using their name for a particular product type. Reputation is a highly dynamic phenomenon needing a sensitive study of theories of probability. The products offered by a company are acceptable by virtue of its high reputation.

2.6.2.4 Word-of-Mouth

Word-of-mouth is a traditional mode of knowledge proliferation. A product liked by a customer receives favorable words for its quality, which proliferates with time. The effect is slow at the beginning but soon increases exponentially, making a quality product less and less asymmetric. A product category, such as fast-food centers and restaurants, has to rely heavily on word-of-mouth. When a product is just launched, word-of-mouth is feeble, but it can grow rapidly and dominate other sources of information as the installed base grows (Sterman 2000).

2.6.2.5 Advertising

Marching toward information revolution, advertising is a lucid way of putting bare facts of product quality across to the customers. A strenuous effort using every available medium helps manufacturers reach their customers. Extensive advertising campaigning is of foremost importance to make information asymmetry subside and to pave the way for a fair market mechanism. Firms can also provide information by direct mail, automatic response system (ARS), or Internet home pages (Park and Kim 2003).

2.6.2.6 Warranty/Guarantee

The important dimension of product quality lies in post-sell services and reliability. Risk-neutral customers may prefer a guaranteed, high-quality product than an unreliable, cheaper one. Sufficient guarantee/warranty induces confidence in a product purchase, which precludes a customer's temptation to purchase a low-quality product. The only harm of this tool is the wanton or reckless use of the product, which results in moral hazard.

Chapter 3
Perceived Quality Through Time

Perceived quality has been studied in several disciplines. But, the research on perceived quality is largely isolated across these disciplines. Literature on marketing science, economics and quality management shed light on perceived quality. Attempts to develop metrics similar to perceived quality have been also made in business management. A few angles to the perceived quality are noticed in the literature.

3.1 Emergence of Perceived Quality

Garvin (1984) identified the eight dimensions of quality. Perceived quality is one of these eight dimensions. Garvin suggested that information content is one of the bases of perceived product quality. He wrote: "Product will be evaluated less on their objective characteristics than on their images, advertising or brand names."

Information as a basis of customers' expected quality has become a part of the studies in economics of information, but its literature has been more related to quality uncertainty due to information asymmetry than to perceived quality. Analysis allowing for the distinction between "objective" and "perceived quality" has not received much attention among economists (Ali and Seshadri 1993).

Perceived quality has received favorable attention in marketing science. Perceived quality is usually defined in the literature as an evaluative judgment of an attitudinal nature (Parasuraman et al. 1988; Carman 1990; Cronin and Taylor 1992; Llusar and Zornoza 2002). Perceived quality is the consumer's judgment about an entity's overall excellence or superiority (Zeithaml 1987). It differs from objective quality (as defined by, for example, Garvin 1983 and Hjorth-Anderson 1984); it is a form of attitude related but not equivalent to satisfaction and results from a comparison of expectations with perceptions of performance (Parasuraman et al. 1988). Garvin (1983), Dodds and Monroe (1985), Hollbrook and Corfman (1985), Jacoby and Olson (1985), and Zeithaml (1987) have emphasized the difference between objective and perceived quality.

L. Wankhade and B. Dabade, *Quality Uncertainty and Perception*,
Contributions to Management Science, DOI 10.1007/978-3-7908-2195-6_3,

A lead article of Steenkamp (1990) uses consumers' psychological underpinning to formulate perceived quality. Customer perceptions on the quality of a firm are observed to be not limited to the evaluation of the characteristics of the product or service but to include all elements that are susceptible to being perceived and evaluated by the client, such as price, image, reputation and so on (Olson and Jacoby 1972; Olson 1977, pp. 267–286; Parasuraman et al. 1988). Besterfield et al. (2003, p. 59) elaborated the customers' perception of quality.

Also, quality may be perceived nationally or internationally. Exporters can use perception's effect to their advantage. As a "halo" that precedes and surrounds goods, this effect actually operates as a brand itself. The country of origin of a product plays a vital role in such national and international perception. Exporters must decide whether they want to double-brand a product with the country of origin or whether they want to suppress identification with the country of origin (Polly 1994).

Perceived quality is now widely viewed as an effective basis for differentiation-based competitive strategies (Aaker and Jacobson 1994; Phillips et al. 1983). Perceived quality is becoming the most essential corporate attribute in the emerging global economy. In the new value equation, perception is a silent partner of the quality factor and can tip the scales (Polly 1994). Brown (1995) coined the term "nation equity" for perceived quality of a product from individual nation's point of view. He observed "Since the products do not start from the same basis in each market, nation equity is likely to play an increasingly pivotal role in quality perception."

Perceived quality and quality perception are the terms used across the literature for the perception of quality. The terms are interchangeable. Mostly, the perceived quality has been the favorable term when discussion involves the assessment of quality on the basis of customer responses.

3.2 Factors of Perceived Quality

A growing body of evidence from the literature suggests various factors of perceived quality:

- *Price* (Scitovszky 1945; Jacoby and Olson 1985; Parasuraman et al. 1985; Sjolander 1992)
- *Market share* (e.g., Phillips et al. 1983; Hellofs and Jacobson 1999)
- *Word-of-mouth* (Brown and Reingen 1987; Reingen and Kernan 1986; Rosen and Olshavsky 1987; Solomon 1986)
- *Brand name* (Andrews and Valenzi 1971; Jacoby et al. 1971)
- *Reputation* (Shapiro 1982; Rogerson 1983; Allen 1984)
- *Financial information* (Aaker and Jacobson 1994)
- *Advertising* (Kirmani 1990, 1997; Abernethy and Butler 1992; Kopalle and Lehmann 1995; Moorthy and Zhao 2000; Tellis and Fornell 1988; Tellis 1988; Tellis et al. 2005; Iyer et al. 2005).

Consumers seem to be forming their quality perceptions through different mechanisms for durable goods and nondurable goods (Moorthy and Zhao 2000). Tellis and Wernerfelt (1987) considered quality and price in the backdrop of asymmetric information. Akerlof (1970) suggested some possible mechanisms to counteract information asymmetry. These are *guarantee*, *brand name reputation*, *chain management* and *licensing practices*. Kopalle and Lehman (2006) examined the optimal advertised quality, actual quality and price for a firm entering a market. Moorman (1998) and Moorman et al. (2005) have related market-side information to competitive strategies. *Supply chain* is not directly involved in quality perception but it is a strong back-end operator. On the basis of a survey, the American Society for Quality ranked important factors of quality perception as (Besterfield et al. 2003, p. 59) performance, features, service, warranty, price and reputation.

3.3 SERVQUAL as a Metric of Perceived Quality

The criteria used by consumers in assessing service quality fit 10 potentially overlapping dimensions. These dimensions are tangibles, reliability, responsiveness, communication, credibility, security, competence, courtesy, understanding/knowing the customer and access (Parasuraman et al. 1985). SERVQUAL is a multiple-item scale for measuring consumer perception of service quality (Parasuraman et al. 1988). The gap between expectations and perceptions form the basis for measurement of the perceived quality of services. Overlapping was reduced and the five dimensions were finalized through the scale purification approach. These are tangibles, reliability, responsiveness, assurance and empathy. The measures about customers' expectation and perception of service quality are developed in a questionnaire for each of the five dimensions. The customers are contacted and responses are obtained on a 7-point scale. Scores are computed as perception minus expectation, and after statistical processing the perceived quality is revealed. Thus the SERVQUAL model became the major approach for perception of service quality.

However, conceptualization and measurement of service quality perception have been the most debated topics in the service marketing literature. This debate continues as is evident from ongoing and largely failed attempts to integrate the SERVQUAL conceptualization into new industries (Brady and Cronin 2001). Brady and Cronin (2001) then revealed a new approach to service quality perception by integrating the "Nordic" perspective (Gronroos 1982, 1984) with the "American" perspective (Parasuraman et al. 1988). Service quality is an abstract and elusive construct because of three features unique to services: intangibility, heterogeneity and inseparability of production and consumption (Parasuraman et al. 1985). In the absence of objective measures, an appropriate approach for assessing the quality of a firm's service is to measure the consumer perceptions of quality (Parasuraman et al. 1988). The SERVQUAL model or integrative approach for service quality perception sees the perception through the consumers' eyes.

3.4 Marketing Science Approach to Product Quality Perception

Most of the literature on perceived quality relates to service quality. Only meager work has been noticed on product quality perception. Mitra and Golder (2006) differentiated perceived quality from objective quality and carried out an extensive empirical research on perceived quality for various products.

According to marketing science literature, perceived quality is determined primarily by objective quality and prior expectations of quality (Boulding et al. 1993, 1999; Olshavsky and Miller 1972; Parasuraman et al. 1985). Figure 3.1 presents the marketing science approach to perceived quality (Mitra and Golder 2006). Objective quality is linked to perceived quality in two ways. The first is the direct contemporaneous link. The second is the indirect lagged link through the updated prior expectations of quality (Bolton and Lemon 1999; Boulding et al. 1993, 1999; Nerlove 1958). This latter link leads to the carryover or delayed effect of objective quality on perceived quality. The gap between prior expectations and objective quality determines the perceived quality.

3.5 Expected Quality Approach from Information Economics

Customers are unable to perceive the quality in the "lemon" situation. Hence, as per the theory of information economics, buyers assume an average price of the offered products. In this situation of averaging, good-quality products are short of the customers' expected quality and lemons are at an advantage over plums. And high-quality products will withdraw from the market. The expected quality will be again re-evaluated by the customers, leading to a few more withdrawals. This mechanism will operate till the market collapses.

But in real world, market does not really collapse. Lemons are picked up by the customers when asymmetry prevails and the average quality is considered as expected quality by the customers. Many illustrations on expected quality are seen in the literature on information economics. But they mostly pertain to the resale market.

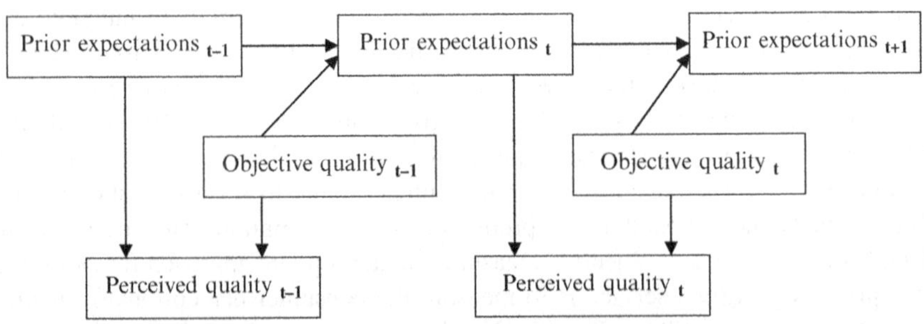

Fig. 3.1 Marketing science framework of perceived quality

Table 3.1 Used car problem

Car	A	B	C	D	E
Quality	0.2	0.4	0.6	0.8	1

3.5.1 *Example: The Market for Used Cars*

Suppose there are five used cars for sale and quality varies from plums to lemons. The quality is represented in Table 3.1. We assume that highest possible quality is 1. The car E is plum and the car A is lemon. The buyer has one-fifth chance of picking a plum or lemon or a quality in between. The sellers of the cars know the exact quality of their cars, whereas the buyers only know the distribution of quality, which is observed to be uniform.

The sellers have decided to sell the quality of "1" at \$2,000. There are many enthusiastic buyers who wish to value a car at \$2,500. The first offered price for the car is \$2,000. All the cars are offered as the maximum price (highest quality) of a car is $\$2{,}000 \times 1 = \$2{,}000$. But the buyers know that the average quality is 0.6 and hence they are willing to pay the price of $\$2{,}500 \times 0.6 = \$1{,}500$. Hence no one will buy a car.

Now, the dealer lowers the offer price to \$1,500. Car D has the price $\$2{,}000 \times 0.8 = \$1{,}600$ and car E has \$2,000. Hence, only cars A, B and C will be offered. The buyers see that the top two cars are not offered. They make a revised estimate. Now the expected average quality is 0.4, and the buyers are willing to pay $\$2{,}500 \times 0.4 = \$1{,}000$. Again no one buys a car.

Thus, the expected quality will gradually decline as the average of the prevailing qualities, and an average price will be offered every time. At no time the cars will be sold and the market crashes.

3.6 Critical Appraisal of Existing Approaches

Perceived quality or quality perception has been largely used in the literature on economics, marketing science and quality management. Specifically, its use has been much larger in marketing science than economics and quality management. Many times the term perceived quality has been replaced by consumer-based terminologies.

The cases from information economics are largely related to the resale market. The expected quality in the resale market is due to the later defects in the product. But the presence of high-quality and low-quality products leads to the quality perception in primary sales. Hence perception analysis for any market and any product type requires a different treatment than the resale market analysis.

Also, the expected quality in information economics has been considered as an average of prevailing qualities. This lacks scientific basis for its analysis. In the

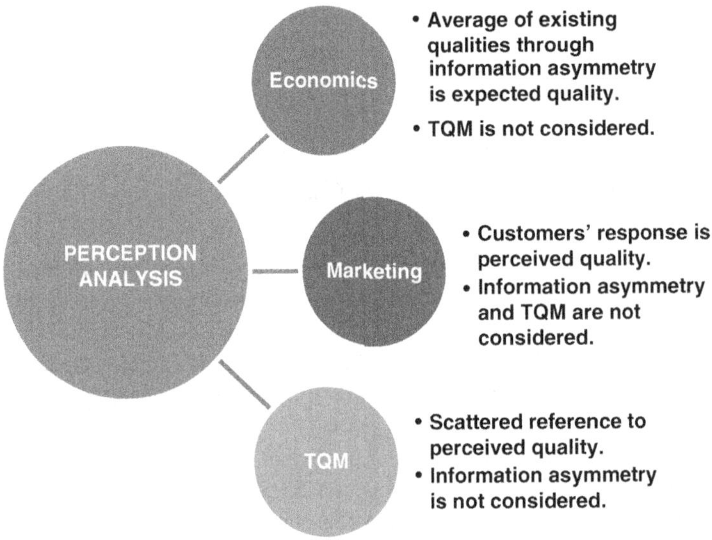

Fig. 3.2 Reflections of the perception process

process of averaging, the expected price continuously decreases and at a point it reaches zero where market collapses. This is not supported by the real market, which never crashes.

The perception analysis has been carried out separately in different disciplines. The reflections of the perception process in different disciplines are shown in Fig. 3.2. It is the expected quality in information economics that uses the knowledge of information asymmetry. Quality perception is normally pursued in marketing science by using the marketing science approach and without the knowledge of information asymmetry. The expected quality from economics and the perceived quality from marketing science should resemble the same force; however, these are becoming different entities.

Perception of product quality in marketing science is based on customer response through sample survey. This method is suitable where items are intangible. The SERVQUAL model uses this method fittingly for service quality. On the contrary, product quality perception is the perception of an objective quality. It can be computed as most of the involved dimensions are measurable at the industry and market end.

Perceived quality in marketing science literature is based on the objective quality and prior expectation of quality. Prior expectation of product quality is a kernel in perceived quality. Customers' expectation of product quality can be true to some extent, but it cannot be the complete truth. There are many products that are manufactured as a part of technology change and considering customer expectations for perceived quality as the only criterion may lead to erroneous results. What then about the all factors that are responsible for perceived quality? The factors of perceived quality gleaned from the literature should have been used to formulate the measure of perceived quality.

Also, customers' expectation of product quality does not have a sound basis for its assumption for product quality, as it has already been considered in quality function deployment (QFD). Prior expectations of product quality are also handled in the continuous process of product development. *Expectation and perception of product quality in marketing science does not offer a control part from the business point of view that can change the process of perception.* Rather, an outcome of the expectation–perception process may indicate outright change in the product which is at par with the new product development.

Similarly, information symmetry plays a crucial role to make product quality the perceived one. Hence information symmetry should be a key component in the process of quality perception. But marketing science conceptual framework on perceived quality lacks the analysis based on presence or absence of information symmetry. *The process of perception of product quality need not evoke an emotional response from the customers but it should measure how far a customer understood the product quality that is manufactured.*

3.7 Need of a Robust Approach

The Net Promoter Score (NPS) has been introduced recently for its simplicity and the way metric links the customers' power of word-of-mouth to the company's growth. A few simple questions regarding the loyalty to a company are asked to customers. An important question that determines the NPS is: "How likely is it that you would recommend our company to a friend or a colleague?" Customers' responses are obtained on 0–10 rating scale. Then customers are grouped on the basis of the ratings. Promoters are based on a 9–10 rating and detractors on a 0–6 rating. The NPS is the ratio of promoters' to detractors' scores. Thus, the NPS makes an interesting assessment of customers' loyalty to a company's business growth. But, the NPS, like any measure of the customer intentions, is inherently unreliable. Now, the renewed urge for another appropriate philosophy still continues.

On the quality management side, it has been suggested that TQM approaches are failing to deliver the anticipated business performance improvements in many companies (Greising 1994; Jacob 1993). The "quality revolution" in the business world and recent academic research has centered on the importance of viewing quality from a customer rather than a supplier perspective (Bounds et al. 1994). The literature points to a strong role for marketing in successful TQM-based strategies (e.g., Cravens et al. 1988), suggesting that interfunctional interactions between marketing and quality may significantly influence the successful formulation and implementation of quality strategies (Morgan and Piercy 1998). However, despite the potential importance of interactions between marketing and quality in affecting quality strategy and business performance outcomes, little conceptual or empirical

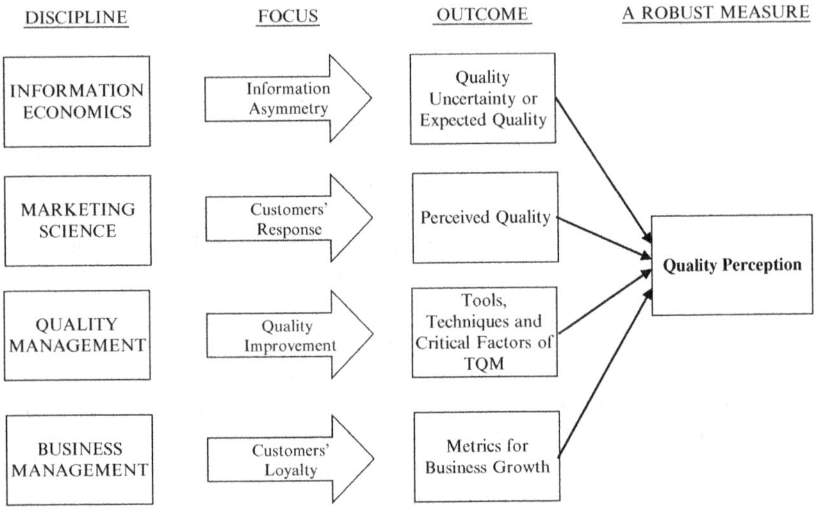

Fig. 3.3 Measure of quality perception

attention has been paid to the interface between the two functions (Day 1994; O'Neal and Lafief 1992).

Perception of quality has seen a continuity of research in the body of literature. Also, the isolated attempts toward the metric formulation have not provided enough strength that could have been possible with a multidisciplinary approach to quality perception. Hence, as shown in Fig. 3.3, the term "Quality Perception" has been coined for the perception of product quality.

Figure 3.3 shows the different approaches that are prominently adopted on quality in various disciplines that have contributed to various outcomes. Information asymmetry from information economics yielded the concept of quality uncertainty and expected quality. Quality management evolved TQM along with the required tools, techniques and critical factors. Business management has a focus on the metrics development which helps business grow. And marketing science predominantly studies customers' response to product or service quality which then becomes perceived quality. The factors for the perception of product quality are also largely similar to the perceived quality with a few exceptions that are specific to the service quality. Another dimension that differentiates quality perception from the earlier notion of perceived quality is that quality perception measure is not only customercentric but also considers the TQM and socioeconomic factors. Business strategies can be developed for the expected business growth by using the concept of quality perception. Thus, multiend formulation of the metric is the most promising direction. The dimension of quality perception that is coined in this book tries to encompass possibly every issue that affects the customer's decision.

Chapter 4
Quality Uncertainty and Quality Perception

The term "quality uncertainty" contains "uncertainty." It implies the use of probability for the analysis. Quality uncertainty (QU) is a failure of product quality at a market end. This is very similar to the product failure that can be studied with failure or hazard rate, failure mode and effect analysis, success tree or failure tree analysis, and like tools of reliability engineering. Thus, the background of probability and reliability engineering is essential for the modeling of QU. Fault tree diagram from reliability engineering is used to model QU. Then basic model of QU is extended to the general model of QU at market end. The term quality perception (QP) has been coined with respect to QU. Success tree diagram is used to model QP. Then, basic model of QP is extended to the general model of QP at market end. The concept of QU has been related to QP by a suitable equation. Behaviors of QU and QPs are studied. Also, QP is analyzed along with product life cycle and information asymmetry.

4.1 Theoretical Background of Probability and Reliability Engineering

Experiment, sample space, and event are the key terms used in probability.

Sample space: The set of all possible outcomes of an experiment is a sample space. Each elementary outcome is represented by a sample point. S is a sample point in Fig. 4.1.

Event: Something that happens that is true or false or possible is an event. A collection of sample points is an event. A is an event in Fig. 4.1.

Complement of A ($\bar{A}$): The set of all elements that do not belong to A. Or the events are complimentary if they cannot occur at the same time. $\bar{A}$ is compliment of A in Fig. 4.1.

Probability is a measurable quantity describing the frequency of occurrence of any given event. Probability allows us to quantify variability in the outcome of any

L. Wankhade and B. Dabade, *Quality Uncertainty and Perception*,
Contributions to Management Science, DOI 10.1007/978-3-7908-2195-6_4,

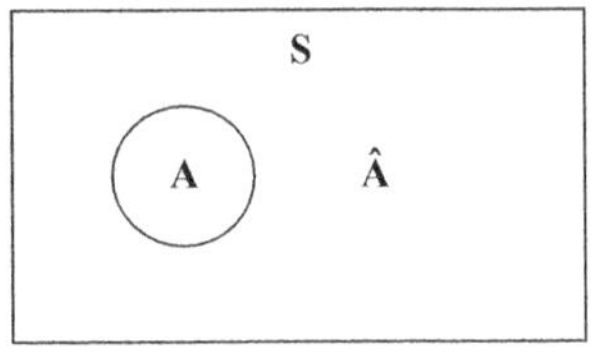

Fig. 4.1 Terms used in probability experiments

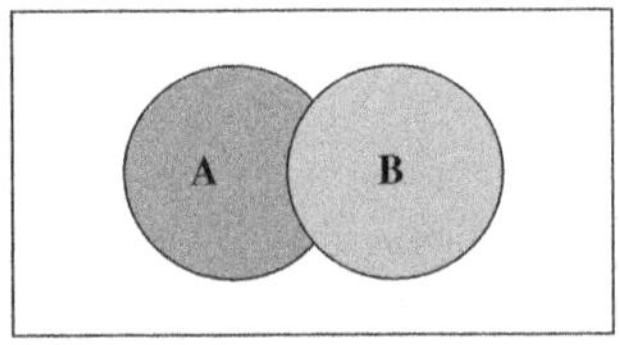

Fig. 4.2 Union of A and B

experiment whose exact outcome cannot be predicted with certainty. The probability of an event is the proportion of times the event would occur in a long run of repeated experiments.

Probability used to be normally explained by an example of "balls in the bag" experiment in our school days. For example, a bag contains 50 red balls, 40 white ones, and 10 black ones. If a boy puts his hand in the bag and picks a ball, what is the probability that it is black? The probability obtained here is 0.1. Hence, the theories of probability help us know the amount of certainty or uncertainty.

Union of A *and* B (A ∪ B): The set of elements that belong to A or B (or both). When two events are independent, the probability of both occurs (Fig. 4.2):

$$P(A \text{ or } B) = P(A) + P(B) - P(A) \times P(B).$$

If we flip a coin and roll a six-sided die, what is the probability that the coin comes up tails or the die comes up 3? Again the probability of a tail is 1/2, and the probability of the die coming up 3 is 1/6.

P (coin comes up tails or die comes up 3) = 1/2 + 1/6 – (1/2 × 1/6) = 7/12.

Intersection of A *and* B (A ∩ B): The set of elements that belong to both A and B. It is also called an intersection of A and B. When two events are independent, the probability of either A or B occurs (Fig. 4.3):

$$P(A \text{ and } B) = P(A) \times P(B).$$

If we flip a coin and roll a six-sided die, what is the probability that the coin comes up tails and the die comes up 3? The two events are independent. The probability of a tail is 1/2. The probability of the die coming up 3 is 1/6.

$$P(\text{coin comes up tails and die comes up } 3) = 1/2 \times 1/6 = 1/12.$$

Fig. 4.3 Intersection of A and B

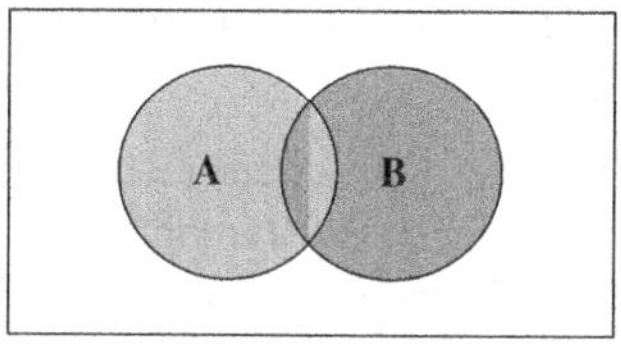

4.2 System Reliability

The product is meant for some or other use. Users want that the product should work for the purpose and should last till its stated life is over. How far the product will last without failure is the life of the product. Here comes the time domain for defining the reliability of the product. Reliability engineering is concerned with failures in the time domain. Product success or failure during the life time of the product is an uncertain phenomenon, but not so much uncertain that it should all the way be given to the vagaries of randomness. On the basis of the parameters of production and operation, essentially, the uncertainty of product life can be explained by the use of scientific analysis using the theories of probability. Thus, reliability is defined as the probability that a product will perform a required function without failure under stated conditions for a stated period of time.

4.2.1 System Reliability with Components in a System

Product consists of many parts, or system consists of many components. Product may fail if any part of the product fails or system fails if any component fails. Each component of a system has its own failure rate or hazard rate. As the system is made up of many components, its failure or hazard should be represented collectively. Hence, the failure of any system is a function of hazard rate of many components.

Components in any particular system can be arranged differently to vary the system's hazard function. For example, think of an electric circuit with a resister or capacitor. As soon as the resister or capacitor fails, the circuit may fail to serve its purpose. If there are many resisters or capacitors making the required value of the resistance or capacitance and if anyone fails, the circuit still works but the resistance or capacitance value will be different and the circuit will work with the changes in current and voltage. If component is supported with a standby component, the circuit works as per the requirements even if the first component fails.

The above example explains the system of a circuit with alternative arrangement along with a change in hazard rate. This implies that a system can be designed for lesser failure rate over a time. The way the system works toward the required reliability is drawn with components' arrangement, and is called reliability block diagram (RBD). In RBD, components are arranged in parallel or series or a mix of both.

4.2.2 Items in Series

Failure of any item results in failure of a system, when items are in series configuration in RBD. It implies that all the items or components must work successfully so that the system should work for the slated period of time. Here reliability of each item is considered for the computation of system reliability. Practically, it is never possible to have the components or items of same reliability measure and they usually differ. Applying the concept of probability, if the system has three items, say all of 70% reliability for slated time, the system reliability will be only 34.30%. Hence, it means that adding components in a series results in debilitating the system reliability. Figure 4.4 shows the generalized RBD for series configuration where $R_1(t)$, $R_2(t)$, and $R_3(t)$ are the reliabilities of items 1, 2, and 3 and R_S is system reliability for time t.

$$R_S(t) = R_1(t) \times R_2(t) \times R_3(t) \times \cdots \times R_N(t).$$

4.2.3 Items in Parallel

In parallel configuration, addition of any item improves the reliability of the system. The system does not fail till all the items in the system fail. Hence, a single item of strong reliability is sufficient to guarantee the success of the system, and added items improve the reliability by probabilistic way. Figure 4.5 shows the generalized RBD for parallel configuration where $R_1(t)$, $R_2(t)$, and $R_3(t)$ are the reliabilities of item 1, 2, and 3 and R_S is system reliability for time t.

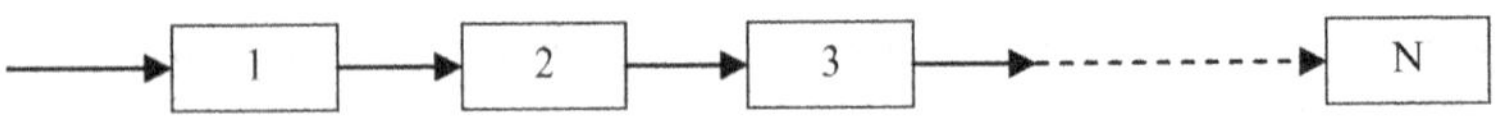

Fig. 4.4 Series configuration

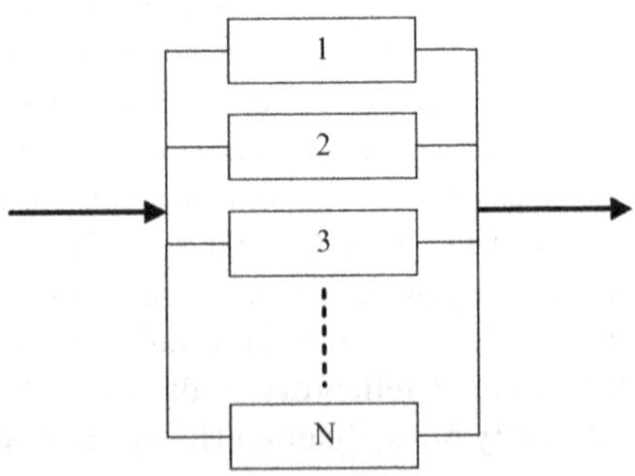

Fig. 4.5 Parallel configuration

If $F(t)$ is a failure of an item, then for parallel configuration,

$$F_S(t) = F_1(t) \times F_2(t) \times F_3(t) \times \cdots \times F_N(t)$$

Since $R_i(t) = 1 - F_i(t)$

$$R_S(t) = 1 - F_S(t)$$

$$R_S(t) = 1 - [1 - R_1(t)] \times [1 - R_2(t)] \times [1 - R_3(t)] \times \cdots \times [1 - R_N(t)].$$

For the same example of a system with three items, say all of 70% reliability for slated time, the system reliability will be 97.30%.

4.3 Success Tree and Failure Tree Method

Success tree and fault tree methods are reliability analysis techniques. The event under consideration is called a top event. The top event is a desirable event in the success tree diagram and an undesirable one in the failure tree diagram. After the top event is drawn, the causes are deduced and the tree goes on branching downward as shown in Fig. 4.6.

The analysis is represented diagrammatically and relationships are made explicit by using gates and symbols. A rectangle exhibits the top event or an event that needs further causal exploration. The causes of the top event are also represented by rectangles, which are explored further. Logic gates are used to link the causes of the events in a tree diagram. Deducing step by step, one can reach to the basic causes at the bottom. The basic cause is represented by a circle. Every subtle cause is not included whereas major causes are identified. The causal exploration of any event will stop when the end causes are drawn by using circles.

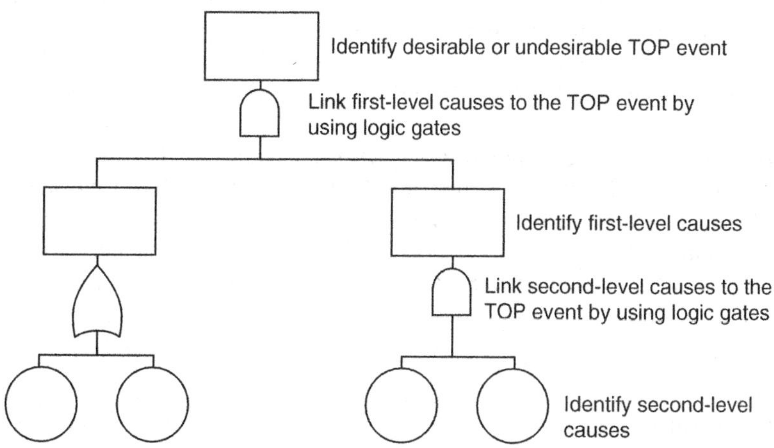

Fig. 4.6 Failure tree or success tree method

The probabilities are applied then to the causes. The success tree or failure tree is constructed by the top to down approach, whereas probabilities are computed by using down to top approach. For example, if A and B are the basic causes of C, with probabilities of 0.3 and 0.6, which are connected by AND gate to C, then the probability of $C = 0.3 \times 0.6 = 0.18$.

Any top event can be represented both ways as success as well as failure. And the difference will be observed in the mechanics of logic gates with some reversion of parameters. Normally, failure tree analysis is preferred in reliability analysis. Usually, any failure analysis is possible. It may cover from simple example of electric switch fault to the failure of satellite launching.

4.4 Paradigm for Mathematical Modeling

Broadly, quality paradigm can be viewed as endogenous and exogenous. Endogenous refers to an industry end where product is manufactured. This is an origin of product quality. Currently, TQM implementation is a prominent determinant of product quality. Tar´ (2005) revealed TQM and its components as follows:

- The critical factors of TQM differ from one author to another, although there are common issues.
- In practice, firms may follow known, accepted, standard models as a guide to carry out quality management.
- TQM is much more than a number of critical factors; it also includes other components, such as tools and techniques of quality improvement.

As seen above, TQM covers critical factors and tools and techniques of quality improvement. But, earlier tools and techniques have continuity in the process of quality control and management and they have become part and parcel of quality improvement programs. We assume that TQM critical factors will be implemented only after the people are adept at earlier tools and techniques of quality improvement. Hence, TQM implementation can be represented by the critical factors.

Total quality management index (TQMI) is an average of TQM constructs that represent the TQM status of a company. Motwani (2001) collated the studies on TQM critical factors and identified the commonalities. The following seven critical factors from the lists of various authors (Ahire et al. 1996; Black and Porter 1996; Flynn et al. 1994; Powell 1995; Saraph et al. 1989; Zeitz et al. 1997) are the mostly accepted constructs:

1. Top management commitment
2. Quality measurement and benchmarking
3. Process management
4. Product design
5. Employee training and empowerment

6. Supplier quality management
7. Customer involvement and satisfaction

TQMI of any industry depends on the implementation of these constructs (critical factors). However, the constructs by themselves are not directly implementable. The critical factors are implemented through various performance measures. These measures that are discussed and tabulated by Motwani (2001) are given in Table 4.1.

These factors form the basis for our modeling task. Similarly, the absence of TQM should be defined as a failure probability. And lack of TQM implementation can also be studied for each measure and ultimately for TQM constructs. We are defining it as a TQM lack.

Similarly, exogenous part plays an equal role to decide whether the product quality will be acceptable or not. A product moves from industry to market where the extent of quality acceptance is measured. Accordingly, acceptance of any product in a market has to be decomposed to major factors. Factors responsible for quality acceptance or nonacceptance in a market are enlisted in Table 4.2. TQM constructs form the endogenous part of the system, whereas market parameters and socio-economic conditions make the exogenous part.

However, if buyers possess the information of product quality, they may either accept or reject it depending on whether quality is good or bad. Thus, it depends upon the quality at origin which is a quality of the manufactured product. And TQM is a suitable measure to universally represent the product quality. Higher TQMI will promise better quality product whereas lower TQMI will lead to lower quality of the product. It implies that QU at the first place will be determined by the quality that is manufactured in industry, which in turn is determined by the level of TQM implementation in the industry.

The amount of information about product quality possessed by customers will determine the chance of acceptance or nonacceptance of the product. Customers possessing the information are information symmetric customers. They perceive the true quality of the product. However, the customers who do not possess the required information are information asymmetric customers. They are unable to perceive the true quality of the product resulting into QU. Hence, information asymmetry produces QU at the second stage. Thus, the amount of asymmetry and symmetry of the product quality information determines the QU and perception. This is quality acceptance or nonacceptance behavior at the market end as shown in Fig. 4.7.

Asymmetry exists in primary sales market in many developing nations where low-quality products are sold by exploiting the asymmetric notion of the people. Prevailing information asymmetry with neutral connotation, obtained irrespective of quality stature, is a general information asymmetry. This type of information asymmetry is observed due to socio-economic and cultural conditions. The first subset of factors from the set of factors generates the general information asymmetry. Another kind of asymmetry is product information asymmetry. General information asymmetry partially stems product information asymmetry. The remaining

Table 4.1 TQM critical factors and performance measures

Critical factor	Meaning of critical factor	Performance measures
Top management commitment	The top management commitment means the involvement of management in creating TQM environment and its implementation. They should be able to motivate the employees to make enhanced efforts to achieve higher quality levels. Management should have vision and support for TQM implementation	Allocating budgets resources Control through visibility Monitoring progress Planning for change
Quality measurement and benchmarking	It incorporates total quality measurement and benchmarking plan of the company. Many authors suggest the use of "zero defects" or "do it right the first time." Defective parts should be in the acceptable zone or minimized to null target as far as possible. The techniques like SQC can be used for measurement	Zero-defects conformance Use SPC for process control Cost of quality Proportion of defects Percentage of products needing rework Defective rate relative to competitors
Process management	This involves value addition to the processes along with improved productivity of workers. For better process management, continuous improvement is required. And philosophies like Kaizen should be implemented	Unit cost Production goals Reduce material handling Design for manufacturability Reduce cycle time Reduce setup time Productivity = finished goods/no. of people or production hours Productivity = total process time/total delivery time
Product design	Sense of better quality should emanate at the product design stage only. Design for quality is an appropriate goal. Many choices are available till the design is finalized. Choices should be explored and selected with customer use and fitness for the purpose in mind	Number of new products introduced Time taken from design to first sale Fitness of use Design quality
Employee training and empowerment	Goals are to be achieved with proper orientation. Many a time, change in mind-sets is required. This is feasible by the use of proper training and orientation programs. Content should include topics from statistical quality control to organizational skills. Seemingly, this component is nonproductive, but it invests in human skills and hence sufficient budget allocation is essential	Training employees Training management Cross-training employees Training/retraining budget

(*continued*)

Table 4.1 (continued)

Critical factor	Meaning of critical factor	Performance measures
Supplier quality management	Motwani (2001) feels that "many companies now support, at least in theory, the need to work more closely with their suppliers. Partnerships with suppliers have the greatest appeal to most companies due to the shared risks associated with the development of new products. Vendor partnerships should be based on a quality program and accepted documentation of progress toward continuous improvement in quality"	Reduce inventory Supplier relations Number of suppliers Inventory turnover Inventory accuracy Implement Kanban Material cost Material availability
Customer involvement and satisfaction	Customers should be the focus for TQM implementation. Timely feedbacks of the customers form the sound basis for product development as well as maintaining the product quality. Company should be sensitive to customers' complaints and must respond within the least possible time	Delivery dependability Operators involved/value-added labor Customer service training budget Prompt handling of complaints Number or percent of complaints Number or percent of orders that are delivered late Broad distribution channels Number of contacts with customers Consumer surveys Time to respond to questions/complaints Responsive repairs Percentage of repeat business

Table 4.2 Factors for quality acceptance or nonacceptance in a market

Endogenous	Exogenous	
Industry end	Market end	Socio-economic end
1. Top management commitment	1. Supplier quality management	1. National culture
2. Quality measurement	2. Working of supply chain	2. Education
3. Training	3. Information asymmetry	3. Economic development
4. Product/service design	4. Post sell services	4. Standard of living
5. Process management/operating procedures	5. Advertising policy	5. Human development index
6. Customers' satisfaction	6. Guarantee/warranty/rebate, etc.	
7. Employee relations		

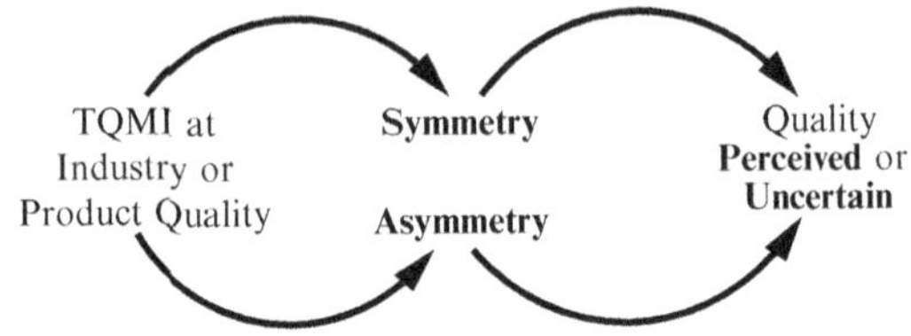

Fig. 4.7 Quality acceptance or nonacceptance at market end

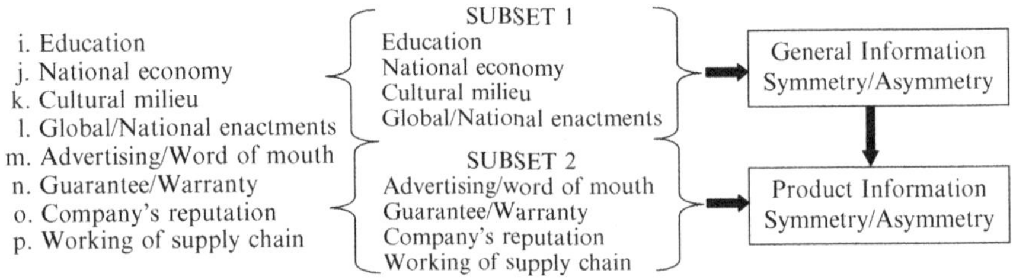

Fig. 4.8 Types of information asymmetry

part depends on the status of marketing strategies or enablers. A company can have the control over product information asymmetry along the product life cycle. Figure 4.8 shows the structure of general and product information asymmetry.

4.5 Behavior of Asymmetry and QU

The subset of socio-economic and cultural conditions is responsible for general information asymmetry, whereas the subset of marketing parameters is responsible for product information asymmetry. The specific aim in the study is to see the decline of these asymmetries. The asymmetry declines along a product life cycle and the time is a major constituent of such behavior. The mathematical formulation of the behavior is also possible.

Decline of the information asymmetry resemble the decay pattern observed in the nature. The formulation of decay is well established and can be applied suitably to information asymmetry. Thus, information asymmetry follows the exponential distribution (decay) at the rate of information proliferation. General information asymmetry could take a long time span to settle at a minimum from the existing one.

Product information asymmetry is a matter of concern for any business performance. Asymmetry is assumed declining along with market growth of the product. And rate of information proliferation determines the asymmetry decline. Required information proliferation rate should be achieved by adopting appropriate marketing enablers or strategies.

Information proliferation rate for general information asymmetry, $\lambda_I = f(i, j, k, l, m, n, o, p)$ can be confined to the market parameters for product information asymmetry. The factors i, j, k, l, m, n, o, p are as enlisted in Fig. 4.8.

$\lambda_{\text{I, for product information asymmetry for developing nations}} = f(i, j, k, l, m, n, o, p)$.

If the socio-economic and cultural conditions are negligible for developed nations, then

$\lambda_{\text{I, for product information asymmetry for developed nations}} = f(m, n, o, p)$.

Product information asymmetry = $e^{-\lambda}I^t$

And, information symmetry = 1 – information asymmetry

Hence, product information symmetry = $1 - e^{-\lambda}I^t$.

Similarly, quality uncertainty also follows exponential decay at the declining rate of information asymmetry.

If TQM lack = 1 – TQM,

$\lambda_{\text{Q For quality uncertainty}} = f(\text{information asymmetry, TQM lack})$.

Similarly, quality uncertainty + quality perception = 1

Quality uncertainty = $e^{-\lambda}Q^t$

Quality perception = $1 - e^{-\lambda}Q^t$

4.6 Fault Tree or Success Tree Analysis

As QU is a failure and QP is a success, fault tree and success tree method is a suitable tool for the analysis. It presents the general framework and is simple for understanding. These analyses can be extended to specific business cases.

In the fault tree, QU is an undesirable top event and hence drawn in a rectangle. Information asymmetry and lack in TQM are the first level causes of QU. These are linked to QU by OR gate. It means that either lack in TQM or information asymmetry makes quality uncertain. The fault tree is constructed in Fig. 4.9.

Further, the second level causes are identified. The lack in TQM is caused due to the nonimplementation of TQM constructs or the lack in implementation of TQM constructs. These are represented by the inverse of TQM constructs. For simplicity, an average of these inverse measures has been taken as a TQM lack.

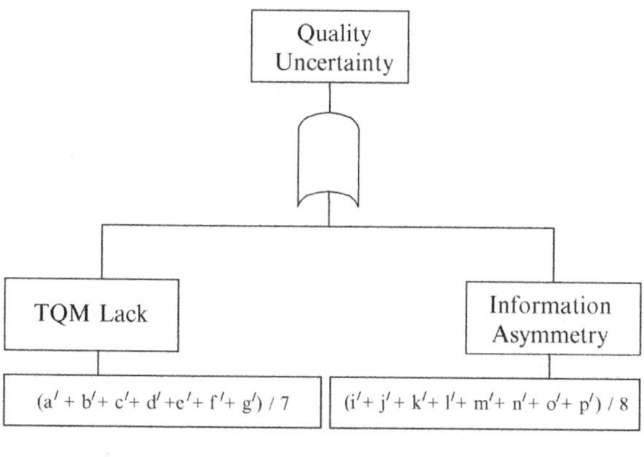

Fig. 4.9 Fault tree diagram of quality uncertainty

But, the further exploration on each TQM construct's lack is also possible. Similarly, information asymmetry results due to the absence of the said factors or lack in the factors. These are taken as an inverse of each factor. Exploration of each factor is also possible.

Inclusion–exclusion principle of set theory offers precise estimate. Mean of all such inversions is more than satisfactory for this general framework for estimation of total lack of TQM. Similarly, mean of inversions of information asymmetry factors results in total information asymmetry at any point of time.

Thus, OR gate is used as a schematic representation connoting that presence of either TQM lack or information asymmetry induces QU. Separately, they constitute the parallel system of reliability engineering. Figure 4.10 shows QU as output for the parallel input of TQM lack and information asymmetry. If the TQM lack at industry end is 0.40 and prevailing information asymmetry measured is 0.30, then QU is 0.58.

$$\text{Quality Uncertainty} = \text{TQM Lack} + \text{Information Asymmetry} - (\text{TQM Lack} \times \text{Information Asymmetry}).$$

A similar explanation is offered for QP by using success tree method. QP is a desirable top event. In Fig. 4.11, information asymmetry is replaced by symmetry

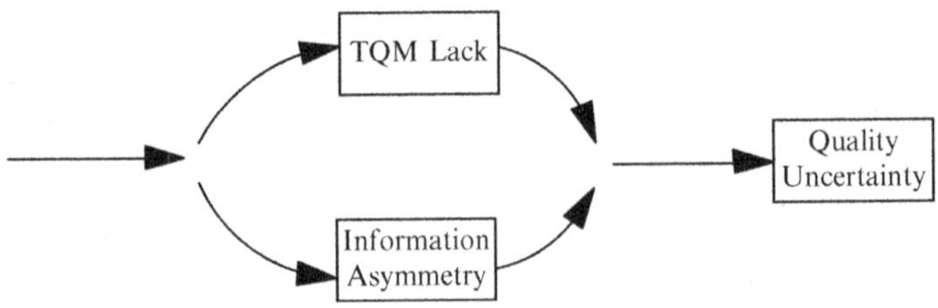

Fig. 4.10 Lack in TQM and information asymmetry in parallel

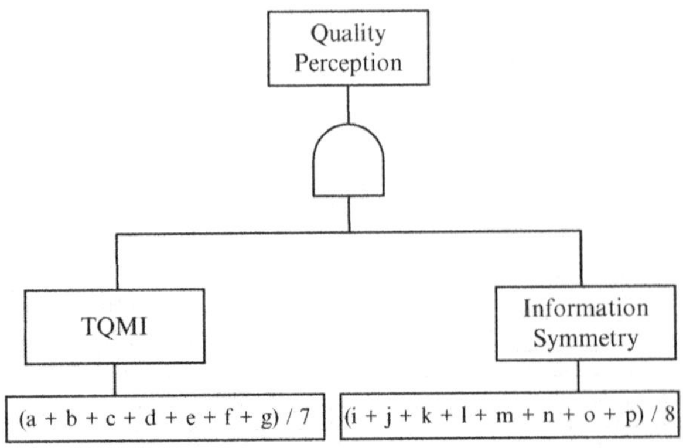

Fig. 4.11 Success tree diagram of quality perception

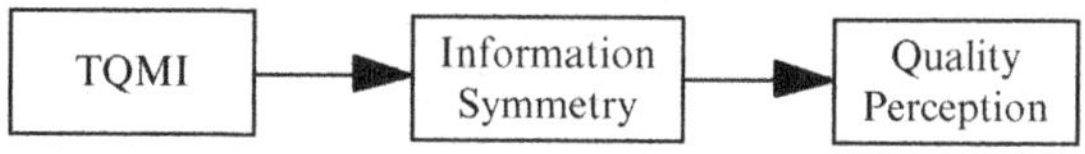

Fig. 4.12 TQMI and information symmetry in series

and QU by QP. TQMI and information symmetry are the first level contributors to QP. These are linked by using AND gate. It means that both TQMI and information symmetry are essential for QP. The implementation of TQM constructs determines TQMI and the level at which the factors are present determines information symmetry as the second level causes.

Separately, they constitute the series system of reliability engineering. Figure 4.12 shows QP as output for the series input of TQMI and information symmetry. If TQM efforts at industry end are 60% and prevailing information symmetry measured is 70%, then QP obtained is 42%. Where,

$$\text{Quality Perception} = \text{TQMI} \times \text{Information Symmetry}$$
$$\text{Quality Perception} + \text{Quality Uncertainty} = 1.$$

Data exploration is the next task to perform in this analysis. TQMI or TQM lack is an industry end entity with seven constructs identified. Nature of these constructs being subjective, they are immeasurable. These are broken down in 45 tangible performance measures and can be computed with data unearthing at industry level. The computations should be moderated to 0–1 scale throughout the process. Harnessing information and communication technologies for data exploration proves to be more efficient and time effective. Data collection is possible by using various reports, sample surveys, and trade volume publications.

4.7 Market with Multiple Players

A series or parallel system of determining QP and QU is relevant to the quality management analysis of only industry or company in a market. This assumes an ideal case and any market is always flooded with a plethora of companies offering similar products. QU and QP in a market will become a conglomerate of all competing products. The product of a particular industry should be seen in the midst of such diverse existence. Hence, total uncertainty or perception should be studied by extending the basic model for many players. Their individual TQM efforts along with prevailing information symmetry or asymmetry will determine the market dynamics for product quality.

The real market has plums and lemons. This is more significant in the Third World. Accordingly, the basic model is extended. Figure 4.13 represents a case of three players, namely, A, B, and C, and depicts QP in parallel along with individually varying product symmetries:

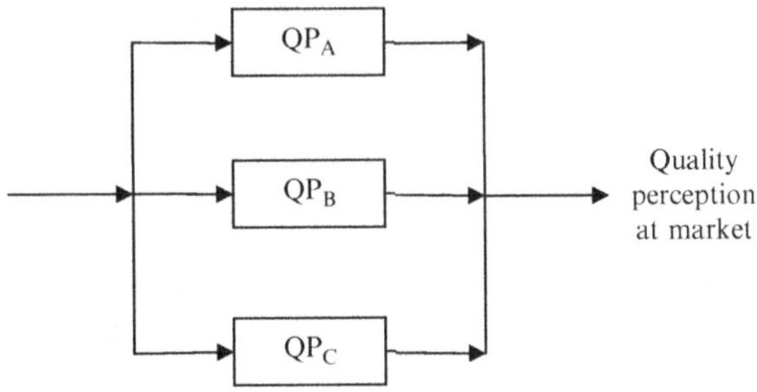

Fig. 4.13 Market quality perception of product type

Fig. 4.14 Market quality uncertainty of product type

$$\text{Quality perception at market end} = QP_A + QP_B + QP_C - QP_A \times QP_B - QP_A \times QP_C - QP_B \times QP_C + QP_A \times QP_B \times QP_C.$$

Similarly, Fig. 4.14 represents QU of the particular product category at market end.

$$\text{Quality uncertainty at market end} = QU_A \times QU_B \times QU_C.$$

Significance of this market analysis lies in the fact that many players offering high quality products alter the market dynamics of QP in a combined way, and individual quality rating could be accessed from a total perspective. It is also true for QU. Many developing nations encounter this hard fact of collective contribution toward perception or uncertainty. Same analysis can also be performed by substituting volume of sales or trade of high or low quality products in place of number of players.

It is an established fact that market is driven by many forces, known as well as hidden, and every analysis is devoid of perfection. The complexities of QU might account for other forces and this analysis can be tuned accordingly. The path set method (normally associated with the success tree) and the cut set method (normally associated with fault tree) should be used to further the analysis toward complexities.

4.8 Computation of Quality Uncertainty and Perception

Analysis in this chapter is based on the theories of reliability engineering. It provides the firm foundation for the otherwise complex issue of QU. Both TQMI and information symmetry are essential for QP, whereas either lack in TQM or information asymmetry will lead to QU.

Table 4.3 Computation of quality perception and quality uncertainty

Case	TQMI	Information symmetry	TQM lack	Information asymmetry	Quality perception	Quality uncertainty
1	1	0	0	1	0	1
2	0	1	1	0	0	1
3	0.5	0.5	0.5	0.5	0.25	0.75
4	0.5	1	0.5	0	0.5	0.5
5	1	1	0	0	1	0

4.8.1 Computation procedure for QU and QP

The example cases in Table 4.3 demonstrate the computation of QU and QP. Computation procedure is explained for case 1.

Illustration of computation of QU and QP for case 1 of Table 4.3

$$\text{Quality Perception} = \text{TQMI} \times \text{Information Symmetry} = 1 \times 0 = 0.$$

$$\begin{aligned}\text{Quality Uncertainty} &= \text{TQM Lack} + \text{Information Asymmetry} - (\text{TQM Lack} \\ &\quad + \text{Information Asymmetry}) \\ &= 0 + 1 - (0 \times 1) = 1.\end{aligned}$$

The case 1 is of very high quality product with TQMI of one. But as information symmetry is zero, the possible QP becomes zero and QU is 100%. Hence, product sees outright rejection. Case 2 is another extreme case where product quality is abysmally low but information symmetry is full. Here perceived quality is zero means that there is altogether absence of quality which can be perceived. Hence, QU is 100% or 1. Case 3 deals with a mediocre quality product in the midst of half information symmetry. Here half of the existing quality is perceived. Case 4 again presents the mediocre quality product but with full of information symmetry. Hence, quality is perceived in true sense as a mediocre quality of 0.5. Case 5 is a special case where the product quality is very high to the extent of 1. Also information symmetry is 100%. The quality must be perceived in a true sense as that of original quality. Hence, QP is 1 or 100% and QU is 0.

4.8.2 Computation procedure for QP at Market End

Similarly, QP or uncertainty at the market end should be computed using the series and parallel concept of reliability engineering. The example cases in Table 4.4 demonstrate the computation of QP at market end for three players.

Table 4.4 Computation of quality perception at market end

Case	A		B		C		Quality perception			Probability of QP
	TQM	Sym	TQM	Sym	TQM	Sym	A	B	C	
1	1	0.2	0.5	0.5	0.5	1	0.2	0.25	0.5	0.7
2	0.4	0.9	1	0.2	0.5	0.5	0.36	0.2	0.25	0.61
3	0.5	0.5	0.7	0.6	0.2	0.7	0.25	0.42	0.14	0.58
4	0.5	1	0.5	0.1	0.5	0.2	0.5	0.05	0.1	0.57
5	1	1	0.9	0.9	0.8	0.9	1	0.81	0.72	1

Illustration of computation of QP at market end for case 1 of Table 4.4

$$\text{Quality Perception} = \text{TQMI} \times \text{Information Symmetry}$$

$$A = 1 \times 0.2 = 0.2$$
$$B = 0.5 \times 0.5 = 0.25$$
$$C = 0.5 \times 1 = 0.5$$

Probability of QP,

$$\begin{aligned}\text{QP} &= \text{QP}_\text{A} + \text{QP}_\text{B} + \text{QP}_\text{C} - \text{QP}_\text{A} \times \text{QP}_\text{B} - \text{QP}_\text{A} \times \text{QP}_\text{C} - \text{QP}_\text{B} \\ &\quad \times \text{QP}_\text{C} + \text{QP}_\text{A} \times \text{QP}_\text{B} \times \text{QP}_\text{C} \\ &= 0.2 + 0.25 + 0.5 - 0.2 \times 0.25 - 0.2 \times 0.5 - 0.25 \times 0.5 \\ &\quad + 0.2 \times 0.25 \times 0.5 \\ &= 0.95 - 0.05 - 0.1 - 0.125 + 0.025 = 0.7\end{aligned}$$

Table 4.4 demonstrates the QP computation of any product type in the market. Last column shows probability of QP of a particular product type in the market. Better quality products improve the probability of QP. If the market has at least a single product with extreme high quality (100% QP) in a product type, then probability of QP of the whole product type becomes 100%. Case 5 supports this assumption. TQMI and information symmetry for product A are extremely high (100%). Hence, QP of A is 100%. QPs of B and C are 81 and 72%. As the QP of A is 100%, the probability of QP of product type (A, B, and C) is 100%. It means, once the highest quality is known, customers can know the other qualities of a product type.

4.9 Product Life Cycle and Information Asymmetry

With the increase in competitiveness and burgeoning market forces after globalization, product life cycle is becoming shorter and shorter. The delay in market entry even by months will make an industry lose sizable amount of product lifetime revenue (Carter and Baker 1992). This is also true for the product quality information. As shown in Fig. 4.15, when information symmetry of a particular product

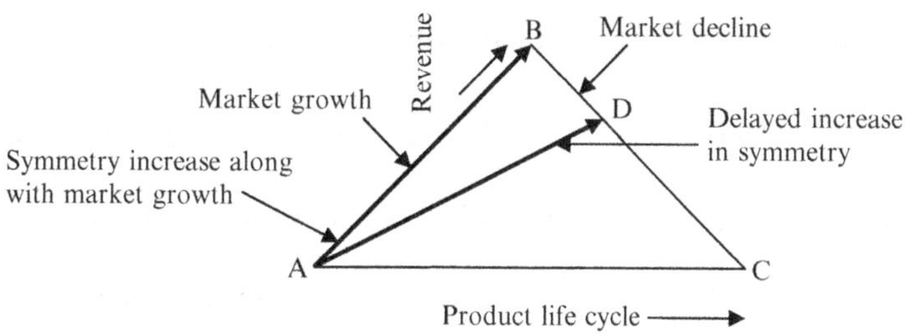

Fig. 4.15 Prevailing symmetry and revenue garnered

Fig. 4.16 Information asymmetry and product life cycle

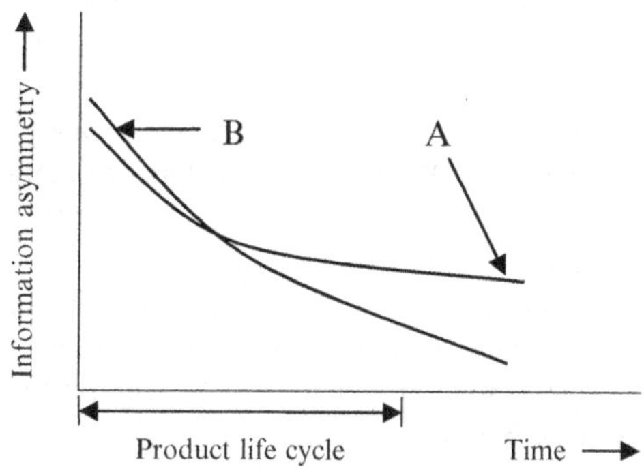

grows along with market growth of the product type (line AB), it is likely that company offering it will garner the business volume as per the predicted market share. The lack or delayed increase in information symmetry (line AD) will cause commensurate revenue loss to the company. The area ABC represents the revenue when information symmetry increases along with the market growth and the area ABD exhibits the loss of revenue because of sluggish rise of symmetry.

Figure 4.16 presents the case of required correlation between product life cycle and information asymmetry. Objective of this illustration is to explain the correlation between information asymmetry decline and product life cycle. The cases are randomly taken. Many more cases can be taken. This type of study indicates which type of information asymmetry decline will be beneficial with respect to the product life cycle. If product life cycle completes earlier than the asymmetry decline, then product may have to undergo a larger range of uncertainty throughout its life cycle.

Case A shows that a product is prone to a wide range of uncertainty and asymmetry saturating throughout its life cycle, and to a higher value of constant uncertainty thenceforth, even if the cycle had been little longer. However, case B presents constant decline of asymmetry with uncertainty always declining and benefits can be foreseen had the product life cycle been longer. Thus, proper correlation of information symmetry/asymmetry of the product quality, juxtaposed to market growth and product life cycle, needs strategic management for better market sustenance.

4.10 Managerial Implications and Strategies

The outcome of this model has wider managerial implications. It implies the contribution of TQMI toward QP. Model connotes the balance (proper mix) required between TQMI at industry end and maneuvering of parameters at market end. It helps in deciding existing market share and probable (planned) market share along with budget allocation for advertising, guarantee/warranty costs, and overall sales promotion. On the basis of this study, some solution strategies are evolved at:

1. To offset the effect of asymmetry of QP, commensurate counteracting institutions such as guarantee, warranty, reputation, rebate, etc., should be employed as a pure strategy. Product quality signaling (like high pricing) may be adopted alongside. Cost benefit analysis can be performed to select the parameters inducing QP. These parameters can be blended and targeted accordingly.
2. The model provides an input to a dynamic customer-relationship management cell. Attempts should be made to strengthen the pool of "quality perception" customers. Modern information tools may be used for information dissipation.
3. Database marketing needs be employed to target customers prone to QU. A larger marketing and sales budget must be earmarked for converting this pool of customers.
4. On the basis of the experiences, grains of symmetry and perception can be incorporated in QFD matrix for product development.
5. Information asymmetry and QP computations must be part and parcel of product life cycle information management tools at industry end. They pave the way for economic product life cycle management.

It is likely to have another thought that QP or uncertainty is in the minds and souls of the customer, and the kind of analysis offered here is more suitable for reliability study of parts, products, or equipments and quite incongruent where customers' emotions and intelligence are mixed together. Though this viewpoint cannot be refuted in totality, this analysis based on the scientific principles of probability has greater extent of validity, for customers' flamboyant thoughts are accommodated in the measure of symmetry or asymmetry of information.

Appendix A

Worksheets 1 and 2

Using Worksheet 1

Worksheet 1 refers to computation of TQMI which depends on the critical factors. Hence, the critical factors are presented in the first column. The status of any critical factor is determined by implementation of the related performance measures. Worksheet 1 when completed presents the TQM implementation status of any company. Entries in worksheet 1 are as follows:

1. The critical factors evolved through the research of the last few decades are given in column 1. These critical factors collectively represent the TQM status.
2. Performance measures for each critical factor are given in column 2.
3. Each performance measure has significance for the TQM implementation in the company.
4. Ponder carefully on each performance measure and fill the value in the third column that is most appropriate to the company. The value should be represented on 0–1 scale in a positive sense.
5. Average the values of performance measures for each critical factor in the last column. This is a value of the critical factor under consideration.
6. Average of the values of all critical factors is TQMI of the company. Write the TQMI at the bottom of the worksheet 1. The value will fall between 0 and 1. If the TQMI value is 0.65, it means that the TQM implementation of the company is around 65%.

Using Worksheet 2

Worksheet 2 refers to computation of information symmetry which also depends on the critical factors. Subset 1 of information asymmetry/symmetry factors represents the socio-economic conditions which are beyond the purview of any company (refer to Fig. 4.8 for subset 1 and 2). Marketing enablers or strategies to improve information symmetry are included in subset 2 of the factors. Subset 2, which makes the major portion of worksheet 2, is crucial in determining information symmetry and is common to developed and developing nations. For companies in developing nations, the values for the factors in subset 1 can be taken in the other issues. For companies in developed nations, the other relevant factors can be taken in other issues. Entries in worksheet 2 are as follows:

1. The identified critical factors are given in column 1. These critical factors collectively represent information symmetry.
2. Performance measures for each critical factor are given in column 2.

3. Each performance measure has significance for improving information symmetry. The performance measures for these do not have universality and may vary from company to company. But a company has to recognize the suitable performance measures for each critical factor. For example, performance measures like complaints about the product, warranty period of the product, and claims in warranty period can be considered for the critical factor of warranty/guarantee.
4. Ponder carefully on each performance measure and fill the value that is most appropriate to the company in the third column. The value should be represented on 0–1 scale.
5. Average the values of performance measures for each critical factor in the last column. This is a value of the critical factor under consideration.
6. Average of the values of all critical factors is information symmetry for the product of the company. Write the information symmetry at the bottom of the worksheet 2. The value will fall between 0 and 1.

After completing worksheet 1 and worksheet 2, compute quality uncertainty and quality perception.

Worksheet 1

Critical factor	Performance measures	Value	Average
Top management commitment	Allocating budgets resources		
	Control through visibility		
	Monitoring progress		
	Planning for change		
Quality measurement and benchmarking	Zero-defects conformance		
	Use SPC for process control		
	Proportion of defects		
	Cost of quality		
	Percentage of products needing rework		
	Defective rate relative to competitors		
Process management	Unit cost		
	Production goals		
	Reduce material handling		
	Design for manufacturability		
	Reduce cycle time		
	Reduce setup time		
	Productivity = finished goods/no. of people or production hours		
	Productivity = total process time/total delivery time		
Product design	Number of new products introduced		
	Time taken from design to first sale		
	Fitness of use		
	Design quality		

(*continued*)

Worksheet 1 (continued)

Critical factor	Performance measures	Value	Average
Employee training and empowerment	Training employees		
	Training management		
	Cross-training employees		
	Training/retraining budget		
Supplier quality management	Reduce inventory		
	Supplier relations		
	Number of suppliers		
	Inventory turnover		
	Inventory accuracy		
	Implement Kanban		
	Material cost		
	Material availability		
Customer involvement and satisfaction	Delivery dependability		
	Operators involved/value-added labor		
	Customer service training budget		
	Prompt handling of complaints		
	Number or percent of complaints		
	Number or percent of orders that are delivered late		
	Broad distribution channels		
	Number of contacts with customers		
	Consumer surveys		
	Time to respond to questions/complaints		
	Responsive repairs		
	Percentage of repeat business		
TQMI (total quality management index) = $(\sum \text{Average})/7$			

Worksheet 2

Critical factor	Performance measures	Value	Average
Reputation			
Advertising			
Warranty/guarantee			

(*continued*)

Worksheet 2 (continued)

Critical factor	Performance measures	Value	Average
Supply chain management			
Word-of-mouth			
Other issues			
Information symmetry = ($\sum$Average)/number of factors used			

Chapter 5
Root Cause and Failure Analysis of Quality Uncertainty

The root-cause analysis evaluates the roles of various factors in generating quality uncertainty and presents its statistical and probabilistic behavior. Seven management tools are used for the root cause analysis of quality uncertainty. The chapter tries to establish and compare the nature of quality uncertainty in developed and developing nations. Succinctly, a simpler method of minimizing quality uncertainty has been developed. This tool can then be used to minimize quality uncertainty in industry-specific cases after seeking expert opinions on involved issues.

An innovative tool is also presented to perform the failure analysis of quality uncertainty in the absence of any empirical data. The data generated through the interrelationship diagram are then used for the failure analysis of quality uncertainty. The feasibility of this failure analysis is demonstrated by illustrating the example of ball bearing. The issues resulting from this analysis are then prioritized as guiding principles for the decision making to minimize quality uncertainty. Use of the tools from the seven management tools makes it easier for quality management professionals to connect quality uncertainty to quality management.

5.1 Theoretical Background of Seven Management Tools

The Union of Japanese Scientists and Engineers felt the need to develop tools to promote innovations and to communicate information. The team researched and developed seven tools often called the seven management tools. Though all the tools were not new, their compilation and calling them the seven management tools were new in 1976.

5.1.1 Affinity Diagram

An affinity diagram was created in the 1960s by Japanese anthropologist Jiro Kawakita. It organizes large amount of disorganized information into various

L. Wankhade and B. Dabade, *Quality Uncertainty and Perception*,
Contributions to Management Science, DOI 10.1007/978-3-7908-2195-6_5,

groupings by natural relationships. All the issues are first explored and listed and broad issues or ideas are identified. These issues are then aligned to the broad ideas. The affinity diagram thus identifies and organizes the issues by natural relationship. Sometimes, a system may have predefined sectors or identifiable sectors. So, the issues can also be grouped under these sectors for obtaining an affinity diagram.

5.1.2 Interrelationship Diagram (ID)

The ID clarifies the connection between various factors involved in a complex situation. The diagram is charted by putting the issues or factors in individual circles or matrices. Arrows are used to represent the cause-and-effect relationships. Also, a suitable weight is assigned to each factor on the basis of expert opinions. The issue or factor with the arrow pointing away represents the root cause and is termed a driver. The issue or factor with an arrow pointing inward suggests a real issue of concern. The sum of the weights of outgoing or incoming arrows shows the gravity of cause and effect for the given factor. Thus, the ID locates the cause-and-effect relationships of the factors so that the identified key drivers and outcomes resolve the problem.

5.1.3 Tree Diagram

The tree diagram reduces the broad objective into increasing levels of manageable details. The broad objective is the primary goal, and then this goal is broken down into subgoals and refined again till we reach at specific level. One can reach to a specific level from a general level by constructing a tree diagram. Tree diagram takes us to the possible level of details. And the accumulation of remedies from specific to general level provides a complete solution.

5.1.4 Prioritization Matrix

Prioritization matrix is a combination of matrix and tree diagram. It is used to prioritize the items. Weight is attached to each item that determines its relative importance. It is the ranking of items and not the sequencing. Thus, it is a priority based action plan that leads to a step by step solution of the issue under consideration.

5.1.5 Matrix Diagram

Matrix diagram exhibits the relationships between the items. The strengths of identified or arrived issues or tasks are weighted against each other. The matrix is drawn, and key issues are written in rows and columns. Their relationship is shown by proper symbols or words. Matrix diagram, thus, highlights the relationships between the items – as roles, strengths, and measurement.

5.1.6 *Process Decision Program Chart*

Tree diagram that is drawn hierarchically down is a process decision program chart. It is similar to failure mode and effect analysis (FMEA). Risks, possible failures, and the consequences of the failures are identified in the first instance. Then, countermeasures are suggested, and implementation plan is charted down.

5.1.7 *Activity Network Diagram*

The arrow diagram for the activities is drawn for systematic implementation of the project. The dependent tasks are drawn sequentially whereas independent tasks are drawn in parallel. The critical paths and parallel paths are identified. Similar to PERT and CPM, attention is given to the critical activities. Activity network diagram helps carry out the project implementation on scheduled time.

5.2 Cause-and-Effect Methodology

Cause-and-effect diagram (CED), current reality tree (CRT), and ID are the popularly used tools for analyzing the root causes of complex problems. Doggett (2004) carried out a comparative study of CED, CRT, and ID. Doggett observed, "CED and ID can be used with little formal training whereas CRT requires comprehensive instruction." Both the CED and the ID exhibited high technical accuracy, whereas the CRT showed mixed results on the level of technical accuracy. He added, "Groups using the CED were seldom able to identify a specific root cause while the groups using the ID did better."

Hence, the ID is selected for its proven simplicity in learning and execution. In this study, affinity diagram, ID, and tree diagram are constructed for understanding quality uncertainty. The study is carried out as per the process detailed in the flowchart given in Fig. 5.1. The flowchart proposed by Besterfield et al. (2003) has been modified to quantitative and qualitative frameworks, with the quantitative framework being a devised component. Data generation and its use for failure analysis are the new techniques added to the existing research methodology.

5.2.1 *Qualitative Framework*

First, we identify the issues related to quality uncertainty. Then, scrambled issues are clustered to represent the sectors of quality uncertainty as an affinity diagram.

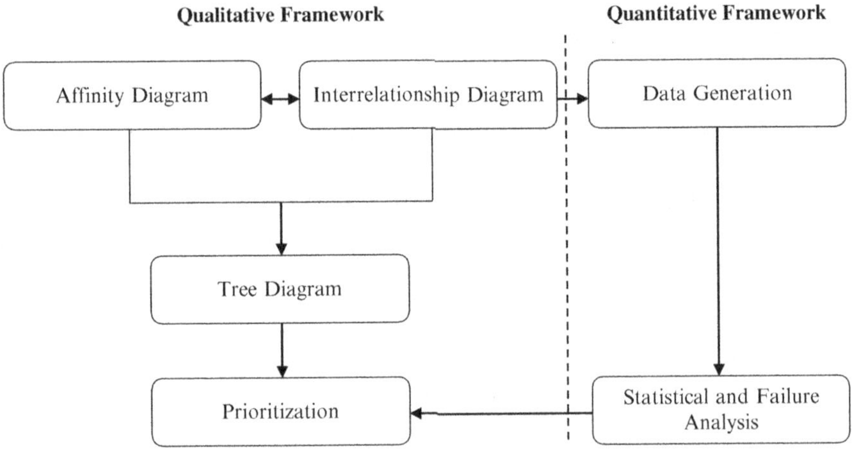

Fig. 5.1 Root cause analysis flowchart

The next step involves constructing an interrelationship matrix wherein the arrows display the cause-and-effect relationships. Then, an analysis is carried out by using the arrow scores arrived at from the ID. Solutions are then offered as objectives and means, in the tree diagram. Finally, the various issues are prioritized for implementation. Normally, the team identifies the issues, and makes the affinity diagram and ID to arrive at a solution for any managerial issue. For the root-cause analysis of quality uncertainty, in this method we have surveyed the causal factors from the literature and have drawn the affinity diagram and the ID.

5.2.2 *Quantitative Framework*

The data inherent in the issues provide better insight. The ID generates the data as represented by the incoming and outgoing arrows. The cause-and-effect-based arrow-count, along with the sum of incoming and outgoing arrows, yields the data without any bias and sampling. Statistics and probability are then used to study the behavior of quality uncertainty. The probability of causing quality uncertainty with respect to different sectors as well as for each factor is determined and then represented in the probability tree. These probability trees are shown in tabular form. The quality uncertainty is analyzed with the help of statistical measures like mean, variance, and standard deviation. Failure analysis is performed on the data generated through the ID. Finally, the possible probability distributions for quality uncertainty are explored.

5.3 An Illustrative Case of Ball Bearing

Use of the ID for failure analysis is demonstrated by considering the case of bearing failure. Bearing is a commonly known part. Hence, it is selected for verifying the use of ID for carrying out failure analysis. Issues associated with bearing failure are identified as shown below:

A. Indentation
B. Overload
C. Contamination
D. Loose fit
E. Fretting
F. Discoloration
G. Misalignment
H. Overheating
I. Vibration
J. Wear
K. Insufficient lubrication
L. Corrosion
M. Electrical pitting
N. Cracking or flaking or spalling
O. Metal to metal contact
P. Improper cleaning
Q. Metal fatigue
R. Improper machining of components

ID is drawn for the listed factors as shown in Fig. 5.2. Each factor is treated at par with a weight of one assigned to each factor. Outgoing arrows are the causes responsible for the failure, whereas the incoming arrows are major outcomes or issues of concern. Incoming arrows can also be treated as failure points. The data

	A	B	C	D	E	F	G	H	I	J	K	L	M	N	O	P	Q	R	In	Out	**Total**
A		←	←		←				↑	↑				↑	←		↑		04	04	08
B								↑	↑	↑	↑			↑			↑		00	07	07
C					↑		↑		↑	↑		↑				←			01	06	07
D							↑		↑									←	01	02	03
E						↑		↑	↑	↑				↑	←				02	06	08
F								←			←	←	←	←					06	00	06
G									↑	↑	↑				↑		↑	←	03	05	08
H											↑	↑	←	↑			↑		03	05	08
I										↑				↑		←	↑	←	08	03	11
J											←		←	↑	←		↑	←	10	02	12
K															↑		↑	←	04	04	08
L													←	↑		←	↑		04	03	07
M														↑		←			01	05	06
N															←		←		10	01	11
O																	↑	←	03	05	08
P																			00	04	04
Q																			09	01	10
R																			00	06	06
Total																			69	69	138

Fig. 5.2 Interrelationship matrix of ball bearing factors

generated on the basis of the incoming arrow-count constitute the general failure data for the part under study. Thus, incoming arrow-count presents the approximate failure mechanism.

Table 5.1 presents the major failures of the bearing, which are obtained by incoming arrow-count from ID. Wear, cracking or flaking or spalling, metal fatigue, and vibration are the known ball bearing failures. These are also revealed by general failure analysis done using ID (see Table 5.1). Similarly, causes of bearing failure can be established. Table 5.2 provides the list of causes obtained through ID analysis. The strongest cause of bearing failure is overload. Contamination, fretting, and improper machining of the components are the other causes responsible for bearing failure. Practical data would prove similar observations.

Also, the general failure data generated through ID should reflect the appropriate probability distribution used in reliability analysis. Weibull distribution is found suitable for the reliability analysis of ball bearings. Incoming arrow-count of ID, also called failure data, is plotted for failure incidence, which is displayed in Fig. 5.3. It suggests that Weibull distribution is suitable for the failure as well as reliability analysis of ball bearings.

The analysis is carried out for general ball bearing failure. Specific circumstances may require reorienting the failure analysis. Weights can be assigned to the

Table 5.1 Failure data of ball bearing

Failure type	Incoming arrow-count	Rank
Wear	10	1
Cracking or flaking or spalling	10	1
Metal fatigue	09	2
Vibration	08	3

Table 5.2 Causes of ball bearing failure

Causes	Outgoing arrow-count	Rank
Overload	07	1
Contamination	06	2
Fretting	06	2
Improper machining of the components	06	2

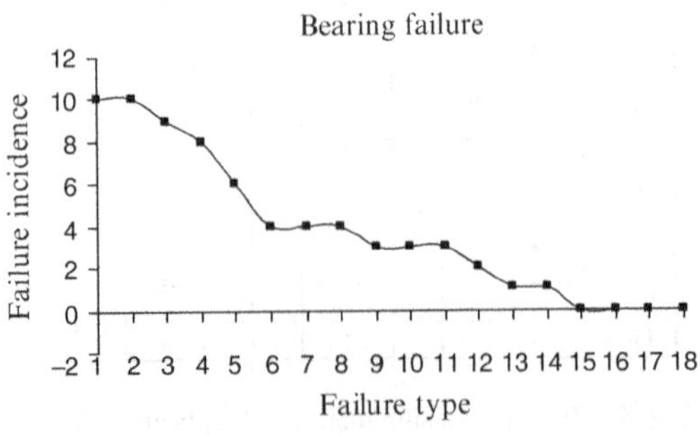

Fig. 5.3 Probability distribution of bearing failure

	A	B	C	E	F	G	H	I	J	L	M	N	O	Q	In	Out	Total
A		←	←					↑	↑			↑	←	↑	2.25	04	6.25
B							↑	↑	↑			↑		↑	00	1.5	1.5
C				↑		↑		↑	↑	↑					00	06	06
E					↑		↑	↑	↑			↑	←		02	05	07
F							←			←	←	←			05	00	05
G								↑	↑				↑	↑	01	04	05
H										↑	←	↑		↑	03	04	07
I									↑			↑		↑	05	03	08
J											←	↑	←	↑	08	02	10
L											←	↑		↑	03	03	06
M												↑			00	05	05
N													←		09	01	10
O															01	04	05
Q															07	00	07
Total															46.25	42.5	88.75

Fig. 5.4 Modified interrelationship matrix of ball bearing factors

arrows on the basis of prevailing conditions. For example, if utmost care is taken for loose fit prevention, lubrication, cleaning, and machining of the components, these factors become nearly null. If the probability of overloading is less, some weight can be assigned to the factor of overload accordingly. Suppose the weight of overloading is reduced from 1 to 0.25. The ID gets modified as shown in Fig. 5.4. Now cracking or flaking or spalling, wear, and metal fatigue are the prominent failures, whereas contamination, fretting, and electrical pitting are the major causes.

5.4 Root-Cause Analysis of Quality Uncertainty

All the major factors determining quality uncertainty are tabulated in Table 5.3. Factors are classified as endogenous and exogenous. Endogenous factors are manageable by the system people while exogenous factors are beyond the purview of quality management. Exogenous factors should be controlled by the state. These factors in Table 5.3 can also be refined further for an industry-specific case.

5.4.1 Affinity Diagram for Quality Uncertainty

Many aspects given in Table 5.3 seem overlapping and repetitive but are retained for the crux of their individuality. For example, guarantee/warranty and post-sell services appear similar, but while guarantee/warranty is impressed on the sale document, the post-sell services, along with other services, are related to the guarantee/warranty realization. Table 5.4 presents affinities of the involved issues in tabular format.

Table 5.3 Factors of quality uncertainty

Endogenous	Exogenous
Information asymmetry	Nonlegislation/enactments by state
Lack in TQM practices	Illiteracy
Inadequate advertising	Poverty
Low reputation of the company	Parochial or narrow cultural values
Deficiency in post-sell services	Poor economy of the nation
Nonenquiry toward customers' feedback	Customers' limited access to information technology tools
Unsatisfactory product performance	Adverse or nonconducive media role
Nonimplementation of QFD	
Seller's biasness	
Insufficient product features	
Feeble supply chain	
Nonstandardization/certification by the company	
Nonguarantee/warranty	
Unaffordable price	
Unattractive brand name or logo	
Unesthetic product	
Unreliable product	

Table 5.4 Affinity diagram

Quality system	Market forces	Information proliferation	Socio-economic conditions
A. Lack in TQM practices	H. Information asymmetry	Q. Inadequate advertising	S. Customers' limited access to information technology tools
B. Nonimplementation of QFD	I. Low reputation of the company	R. Adverse or non-conducive media role	T. Illiteracy
C. Nonstandardization/ certification by company	J. Deficiency in post-sell services		U. Poverty
D. Unsatisfactory product performance	K. Seller's biasness		V. Parochial or narrow cultural values
E. Insufficient product features	L. Nonguarantee/warranty		W. Poor economy of the nation
F. Unreliable product	M. Feeble supply chain		X. Nonlegislation/ enactments by state
G. Unesthetic product	N. Nonenquiry toward customer feedback		
	O. Unaffordable price		
	P. Unattractive brand name or logo		

These issues or factors are fitted to four major categories, viz. quality system, information proliferation, socio-economic conditions, and market forces. The first one, that is, quality system, recognizes the issues on the industry side, which are related to the product and its quality. The second, information proliferation, as it signifies, involves advertising, media, and IT tools. Socio-economic conditions specifically cover the range of issues addressing the customer's position. The last, market sector, has information asymmetry as a prominent factor causing quality uncertainty.

5.4.2 Interrelationship Diagram for Quality Uncertainty

Figure 5.5 exhibits an interrelationship matrix of quality uncertainty constructed utilizing the issues listed in Table 5.4. The issues are the ones appropriate to developing nations. Socio-economic aspects demarcate developed nations from developing nations, as differing socio-economic aspects are observed between the two types of nations. For example, the socio-economic factors from the list considered in this work are presumed nearly absent for developed nations. So the interrelationship matrix developed in Fig. 5.5 is pertinent to developing nations only. As a generic case, the weight assumed for each factor is one. Every counted arrow has the weight count of one. So both weight count and arrow-count have the same meaning for the construction of the interrelationship matrix of quality uncertainty.

The incoming and outgoing arrow-counts in the ID are observed to be 85 each in number, leading to a total of 170. Roughly, half of the factors are causes and the remaining the effects of quality uncertainty. A simple observation of the sum of incoming and outgoing arrow-counts brings out the key players. Low reputation of the company, nonenquiry of customers' feedback, information asymmetry, and lack of TQM practices are dominant players in a descending order, with a total arrow-count (incoming + outgoing) of 16, 13, 12, and 10, respectively.

Further analysis of outgoing and incoming arrow-counts unfolds the key drivers and outcomes, which are listed in Tables 5.5 and 5.6. As seen from Table 5.5, an absence of communication with the customers for obtaining their feedback is observed as the foremost root cause for quality uncertainty. The next three factors are placed at second rank, and out of these two are exogenous. These are the implementation of quality function deployment (QFD) and also the standardization or certification by the company. Both these factors are next in importance from the viewpoint of quality management. Then, sellers' bias, nonguarantee/warranty, and feeble supply chain are some other root causes in the descending order. Illiteracy, poor economy of the nation, parochial cultural values, and nonenactments by the state are the root causes specific to developing nations. From the ten identified key drivers, six factors – with a share of 45% – are endogenous. These should therefore be targeted commensurately.

The major outcomes are ranked in Table 5.6. The prevailing reputation of the company is a prominent outcome with the largest share of nearly 20%. This is followed by information asymmetry. Deficiency in post-sell services, lack of TQM practice, and an unreliable product nature are some of the other outcomes of the root-cause analysis. Out of these, the major outcomes are the issues of real concern and need a further discourse to shed more light on quality uncertainty.

Table 5.7 presents the arrow-count based comparison of the sectors. The incoming, outgoing, and total arrows are counted for quality system, socio-economic sector, information proliferation sector, and market forces. The extent to which each sector influences quality uncertainty has been ranked in Table 5.8. The major factor among these was market sector, which has nearly half the share in causing quality uncertainty. Quality system was the second factor with a significant share of

	A	B	C	D	E	F	G	H	I	J	K	L	M	N	O	P	Q	R	S	T	U	V	W	X	In	Out	Total
A		←	←	↑	↑	↑			↑					←								←	←	←	6	4	10
B				↑	↑	↑	↑	↑	↑					←								←			2	7	09
C				↑		↑		↑	↑	↑													←	←	2	6	08
D						↑			↑					←											4	2	06
E									↑					←											3	1	04
F									↑			←													5	1	06
G														←											2	0	02
H									↑		↑	←		←		↑	←	←	←	←				←	9	3	12
I										←	←	←	←	←	←	←	←	←							16	0	16
J											←	←	←	←						←				←	7	1	08
K												↑	↑	↑						←		←			3	5	08
L													←	↑											2	5	07
M														↑									←		2	4	06
N															↑							←			4	9	13
O																					←		←		3	1	04
P																	←	↑		←					3	2	05
Q																			←	←					2	3	05
R																						←		←	3	2	05
S																				←	←		←		3	2	05
T																						↑	←		1	7	08
U																							←		1	2	03
V																								↑	1	6	07
W																									0	7	07
X																									1	5	06
Total																									85	85	170

Fig. 5.5 Interrelationship matrix of quality uncertainty factors

Table 5.5 Key drivers

Factor	Count	Percentage	Rank
Nonenquiry toward customers' feedback	9	10.58	1
Illiteracy	7	8.24	2
Poor economy of the nation	7	8.24	2
Nonimplementation of QFD	7	8.24	2
Parochial or narrow cultural values	6	7.05	3
Nonstandardization/certification by the company	6	7.05	3
Seller's biasness	5	5.88	4
Nonguarantee/warranty	5	5.88	4
Nonlegislation/enactments by state	5	5.88	4
Feeble supply chain	4	4.70	5

Table 5.6 Major outcomes

Factor	Count	Percentage	Rank
Low reputation of the company	16	18.82	1
Information asymmetry	9	10.58	2
Deficiency in post-sell services	7	8.23	3
Lack in TQM practices	6	7.05	4
Unreliable product	5	5.88	5

Table 5.7 Sector-wise performance (developing nations)

	Quality system				Market forces				Information proliferation				Socio-economic		
	In	Out	Total		In	Out	Total		In	Out	Total		In	Out	Total
A	6	4	10	H	9	3	12	Q	2	3	05	S	3	2	05
B	2	7	09	I	16	0	16	R	3	2	05	T	1	7	08
C	2	6	08	J	7	1	08					U	1	2	03
D	4	2	06	K	3	5	08					V	1	6	07
E	3	1	04	L	2	5	07					W	0	7	07
F	5	1	06	M	2	4	06					X	1	5	06
G	2	0	02	H	4	9	13								
				I	3	1	04								
				J	3	2	05								
Total	24	21	45		49	30	79		5	5	10		7	29	36
%	28.3	24.7	26.4		57.6	35.3	46.4		5.8	5.8	5.8		8.2	34.1	21.2

Table 5.8 Sectors' involvement in quality uncertainty

Sector	Total count	Percentage	Rank
Market forces	79	46.47	1
Quality system	45	26.47	2
Socio-economic conditions	36	21.17	3
Information proliferation	10	5.88	4

27%. Both these sectors are in purview of quality management with the combined share of 73%.

The subtleties of key drivers and the major outcomes from the various sectors are also shown in Table 5.7. The three factors of market forces, socio-economic conditions, and quality system were all equal contributors to quality uncertainty with an individual share range of 25–35%. The effect of quality uncertainty was heavily felt on market mechanism with a huge percentile of 57. This is in concurrence with the established principle of "quality uncertainty and market mechanism." Quality system showed 30% effect on quality uncertainty. As socio-economic conditions are exogenous, both the factors of market forces and quality system can be targeted to curb quality uncertainty.

5.5 Innovative Method of Failure Analysis

5.5.1 *Quality Uncertainty as a Probability*

As the term "quality uncertainty" contains the function of "uncertainty," it implies the use of probability for analysis. The incidence of quality uncertainty means a failure of the product quality at the market end, very similar to product failure. The failure of a product can be studied with the help of tools like the failure or hazard rate, failure mode and effect analysis, and success tree or failure tree analysis. The semblance of treatment seems valid in learning the concept of quality uncertainty. The probability of quality uncertainty is denoted by the symbol P_{qu}.

5.5.2 *Failure Analysis of Quality Uncertainty*

The incoming arrows show the effects or reflections of quality uncertainty, which are more relevant to this study. These incoming arrow-counts can be used as failure points for any part or subsystem or system understudy. Incoming arrow-counts translated into probabilities can thus be the probabilities for failure. A case of ball bearings demonstrates failure analysis using incoming arrow-count data generated through an ID.

Similarly, Table 5.9 shows probabilities of quality uncertainty (P_{qu}) obtained by the incidences of incoming arrows in a tabular form. The sector's or factor's share in causing quality uncertainty has been expressed in percentage or as a fraction on the scale of 0–1. Wear, cracking, flaking, and metal fatigue are all failures of the product in question – ball bearings. Similarly, low reputation of the company, information asymmetry, etc., are failures of product quality. The rest of these factors identified as major outcomes in Table 5.6 comprises the list of failures of product quality.

The probabilities of quality uncertainty in the market, quality management system, information nonproliferation, and socio-economic sectors are 0.5764, 0.2823, 0.0588, and 0.0823, respectively. Probability of quality uncertainty is nearly 60% at the market end, suggesting that equal efforts should be made toward the market side. Lack of a quality system adds a quarter to the quality uncertainty,

Table 5.9 Probability tree of quality uncertainty in developing nations

Sectors	Issues	Factors	P_{qu}
Market forces $P_{qu} = 0.5764$	Reputation through information asymmetry and other factors $P_{qu} = 0.5176$	Reputation	0.188
		Information asymmetry	0.105
		Nonguarantee/warranty	0.023
		Deficient post-sell services	0.082
		Lack of customer feedback	0.047
		Unaffordable price	0.035
		Unattractive brand name or logo	0.035
	Supply chain $P_{qu} = 0.058$	Feeble supply chain	0.023
		Sellers biasness	0.035
Quality system $P_{qu} = 0.2823$	Product development $P_{qu} = 0.1882$	Non-QFD	0.023
		Unreliable product	0.058
		Insufficient features	0.035
		Unsatisfactory performance	0.047
		Unesthetic product	0.023
	TQM practices $P_{qu} = 0.094$	Insufficient TQM implementation	0.070
		Noncertification	0.023
Information nonproliferation $P_{qu} = 0.0588$	Inadequate advertising		0.023
	Nonconducive media role		0.035
Socio-economic $P_{qu} = 0.0823$	Customers limited access to IT tools		0.035
	Illiteracy		0.011
	Poverty		0.011
	Parochial or narrow cultural values		0.011
	Poor economy of the nation		0.000
	Nonlegislation/enactments by state		0.011
Total			$0.99 \approx 1$

also necessitating significant attention. Eighty-five percent of the quality uncertainty is caused by failure in market and quality management system.

Coming to the subgroups, low reputation caused by information asymmetry and other factors (Table 5.9) is the single most prominent quality failure describing half of the quality uncertainty. The failure in proper product development leads to 18% of quality uncertainty. Insufficient TQM practices contribute to 10% of the uncertainty when compared with 6% contributed by feeble supply chain.

Similarly, the probability tree can be obtained for developed nations. Table 5.10 shows the relative probability tree for developed nations. Here, the probabilities of quality uncertainty in the market, quality management system, and information nonproliferation sectors are 0.4352, 0.2217, and 0.0017, respectively.

When failure probabilities of developed nations are compared with those of developing nations, it is observed that the probability of quality uncertainty in developed nations is lesser in the market by 14.12%, in the quality system by 6.06%, in the information sector by 5.71%, and in the socio-economic sector

Table 5.10 Probability tree of quality uncertainty in developed nations

Sectors	Issues	Factors	P_{qu}
Market forces $P_{qu} = 0.4352$	Reputation through information asymmetry and other factors $P_{qu} = 0.4116$	Reputation	0.188
		Information asymmetry	0.070
		Nonguarantee/warranty	0.023
		Deficient post-sell services	0.058
		Lack of customers feedback	0.035
		Unaffordable price	0.011
		Unattractive brand name or logo	0.023
	Supply chain $P_{qu} = 0.023$	Feeble supply chain	0.011
		Sellers' biasness	0.011
Quality system $P_{qu} = 0.2217$	Product development $P_{qu} = 0.1848$	Non-QFD	0.012
		Unreliable product	0.061
		Insufficient features	0.036
		Unsatisfactory performance	0.049
		Unesthetic product	0.024
	TQM practices $P_{qu} = 0.0369$	Insufficient TQM implementation	0.036
		Noncertification	0.000
Information nonproliferation $P_{qu} = 0.0017$	Inadequate advertising		0.000
	Nonconducive media role		0.001
Total			0.65

by 8.23%. Hence, total quality uncertainty in developed nations is roughly 35% lesser than that in developing nations.

5.5.3 *Probability Distributions for Quality Uncertainty*

The next step in the analysis is fitting the statistical data to appropriate probability distribution. Component failure follows a distribution by failure types. The data of incoming arrows are used as failure data for this work. The data for developed nations (Table 5.10) are derived by developing an ID without socio-economic reflections. Measures like range, mean, mode, median, variance, standard deviation, skewness, and kurtosis are calculated for both sets of failure data. These measures are shown in Table 5.11. Statistical measures of distribution suggest a positively skewed distribution with leptokurtosis. The distribution is relatively varied and shows a leftward shift for the developed nations with a difference of 0.0039 in standard deviation and of 0.0046 in mean.

Failure data obtained through the ID are general and representative in nature. Probability distributions, which are more often used in statistical analysis, are preferred for goodness-of-fit tests. Beta, chi-square, exponential, extreme value, gamma, log-normal, normal, Pareto, and Weibull distributions are used for data fitting. A method selection to perform goodness-of-fit test therefore requires statistical acumen.

Table 5.11 Statistical measures of failure data distribution

Type	Range	Mean	Mode	Median	Variance	Standard deviation	Skewness	Kurtosis
Developing	0.188	0.0412	0.023000	0.03500	0.001501	0.0396	2.3652	9.0838
Developed	0.188	0.0366	0.011000	0.02350	0.001794	0.0435	2.4651	9.3485

Table 5.12 Goodness-of-fit tests applied to quality uncertainty in developing nations

	Anderson–Darling	Chi-square	Kolmogorove–Smirnov	Combined score	Final rank
Beta	7	5	4	16	5
Chi-square	6	7	7	20	6
Exponential	4	6	5	15	4
Extreme value	3	1	3	7	3
Gamma	2	2	2	6	2
Log-normal	1	3	1	5	1
Normal	5	4	6	15	4
Pareto	Invalid	Invalid	Invalid	Invalid	Invalid
Weibull	Invalid	Invalid	Invalid	Invalid	Invalid

Connor et al. stated, "A number of methods are available to test how closely a set of data fit an assumed distribution. As with s-significance testing, the power of these tests in rejecting incorrect hypotheses varies with number and the type of data available, and with the assumption being tested" (Connor et al. 2002).

Chi-square, Kolmogorove–Smirnov, and Anderson–Darling tests are carried out separately, and the obtained ranks are combined to arrive at a final rank. The test results of quality uncertainty in developing nations are shown in Table 5.12, and the test results of quality uncertainty in developed nations are shown in Table 5.13. The generated data are not precise and lack completeness. Distributions obtained by the generated data provide guidelines and are of suggestive nature.

From Table 5.12, it can be observed that the log-normal, gamma, and extreme value distributions fit to the quality uncertainty data of developing nations in descending order, with a score of 5, 6, and 7. The next possible fits are normal and exponential, with a score of 15 each. However, the gap between third and fourth rank is a sizeable one, and hence density function from the fourth rank onward should be rejected.

An insignificant difference is noticed among the top three ranks. So, it becomes logical to study quality uncertainty with log-normal, gamma, and extreme value distributions. Exponential distribution (Table 5.13) has been observed as the most optimum for the data concerning developed nations.

5.5.4 Quality Uncertainty Distribution in Developing Nations

Anderson–Darling and Kolmogorove–Smirnov tests suggest that the best fit is that of log-normal distribution for looking at the quality uncertainty data of developing nations. Log-normal distribution is such that the data translated into natural log

Table 5.13 Goodness-of-fit tests applied to quality uncertainty in developed nations

	Anderson–Darling	Chi-square	Kolmogorove–Smirnov	Combined score	Final rank
Beta	5	2	4	11	4
Chi-square	4	5	5	14	5
Exponential	1	1	1	3	1
Extreme value	2	3	2	7	2
Gamma	Invalid	Invalid	Invalid	Invalid	Invalid
Log-normal	Invalid	Invalid	Invalid	Invalid	Invalid
Normal	3	4	3	10	3
Pareto	Invalid	Invalid	Invalid	Invalid	Invalid
Weibull	Invalid	Invalid	Invalid	Invalid	Invalid

represent a normal distribution. The model is particularly suitable for failure processes that are the result of many small multiplicative errors (Gupta 1996). Quality uncertainty in developing nations particularly depends on many interactive factors, and hence log-normal distribution is the most appropriate one. Figure 5.6 shows the best fit of log-normal distribution for quality uncertainty in developing nations with a mean of 0.0488884 and a standard deviation of 0.035764.

Equally valid and relevant is a gamma distribution, which was the second preference shown by all tests. Many distributions like exponential, Erlang, and chi-square are derived from the gamma distribution that offers a wide array of applications in reliability engineering. Gamma distribution is the sum of one or more exponentially distributed variables. Figure 5.7 shows the gamma distribution for quality uncertainty in developing nations with a shape value of 1.5532 and a scale value of 0.027391.

Chi-square goodness-of-fit tests indicate extreme value distribution as the most suitable for quality uncertainty in developing nations. The extreme value distribution models component failure due to extreme phenomenon (Gupta 1996). This is also called the "weakest link" type failure. Low reputation, information asymmetry, deficiency in post-sell services, and insufficient TQM implementation are the weak links making quality an uncertain phenomenon. These factors collectively yield extreme value distribution. Figure 5.8 shows the extreme value distribution for quality uncertainty in developing nations with a shape value of 0.026610 and a scale value of 0.021812.

5.5.5 *Quality Uncertainty Distribution in Developed Nations*

All the tests (Table 5.13) imply that the exponential distribution is the fittest for quality uncertainty in developed nations. The exponential distribution exhibits a simple, constant, hazard rate model, a situation that is often realistic (Gupta 1996). Minimization of uncertainty due to otherwise extreme factors like asymmetry, TQM content, etc. ensues in quality uncertainty at a constant rate in developed

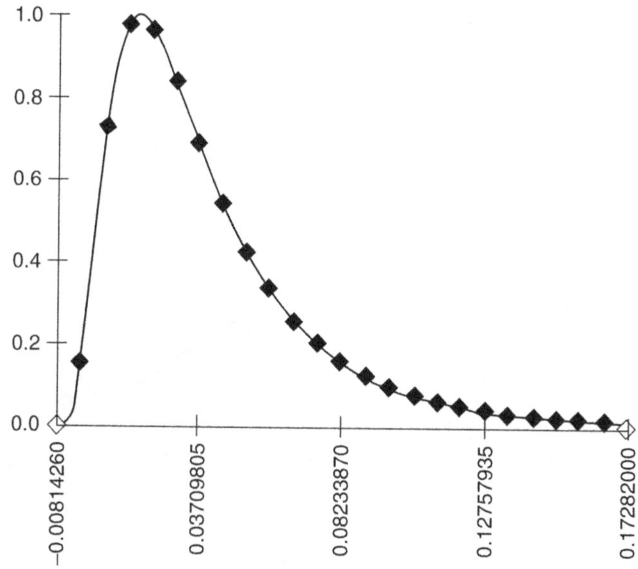

Fig. 5.6 Log-normal distribution

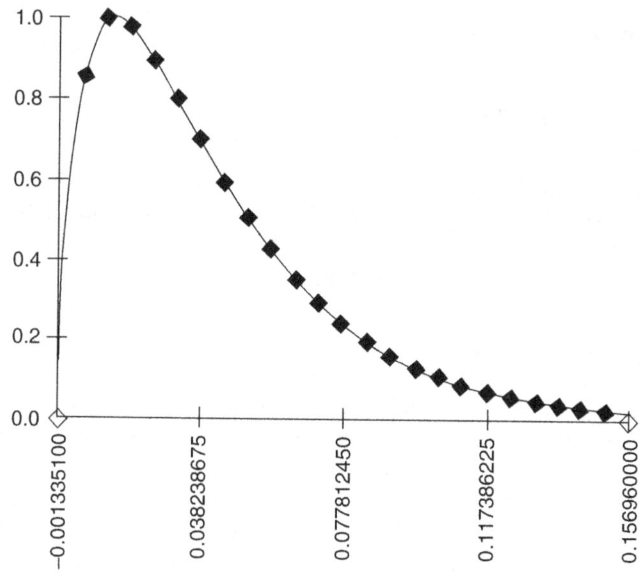

Fig. 5.7 Gamma distribution

nations. Values of the factors causing quality uncertainty gradually descend to make exponential distribution a suitable function. Figure 5.9 shows an exponential distribution for quality uncertainty in developed nations with $\lambda = 0.036056$.

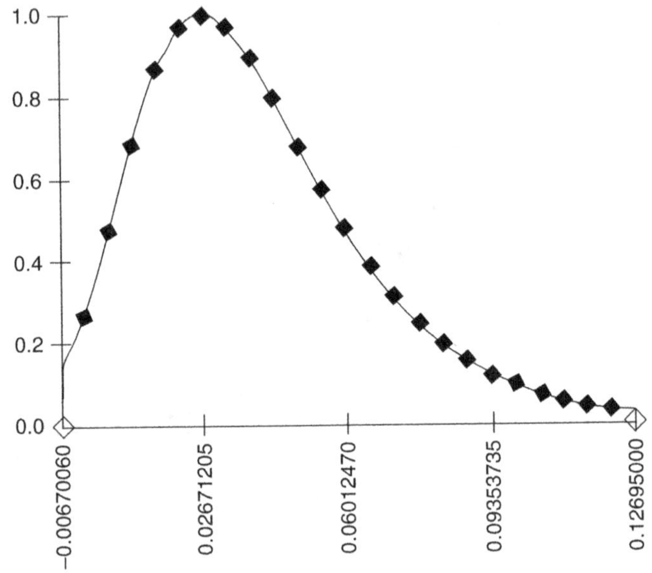

Fig. 5.8 Extreme value distribution

5.6 Generic Solutions

5.6.1 Tree Diagram

Tree diagram is a qualitative solution obtained through the analysis of ID and is a kind of road map drawn to minimize the quality uncertainty. Although managerial acumen is still required for its execution, tree diagram reduces broad objectives into increasing levels of manageable detail. Endogenous factors are considered for the construction of tree diagram. Exogenous factors are left out as they are beyond the scope of quality management. Minimizing quality uncertainty is an objective of the tree diagram as shown in Fig. 5.10. Quality uncertainty is then crystallized into five goals at the second level. Improving the market mechanism, quality system, product reliability, product performance, and features are various steps toward attaining this objective. These are broken down into further details.

Higher reliability, better product performance, and more features can be achieved by concurrent product development. This requires a collaborative effort of experts from the product development team. Customer feedbacks are translated into a QFD matrix. Technical outputs translated from QFD matrix should then be discussed in all perspectives by the concurrent team. The quality system should be improved by implementing the TQM philosophy. This requires the implementation of maximum TQM constructs.

Standardization or certification should be carried out for further quality enhancement. Market mechanism is a kernel of quality uncertainty that needs the highest attention. Both company reputation and information symmetry are prominent

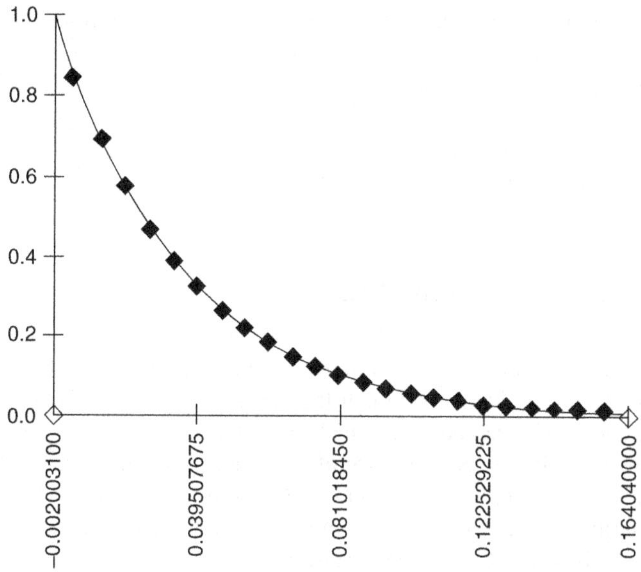

Fig. 5.9 Exponential distribution

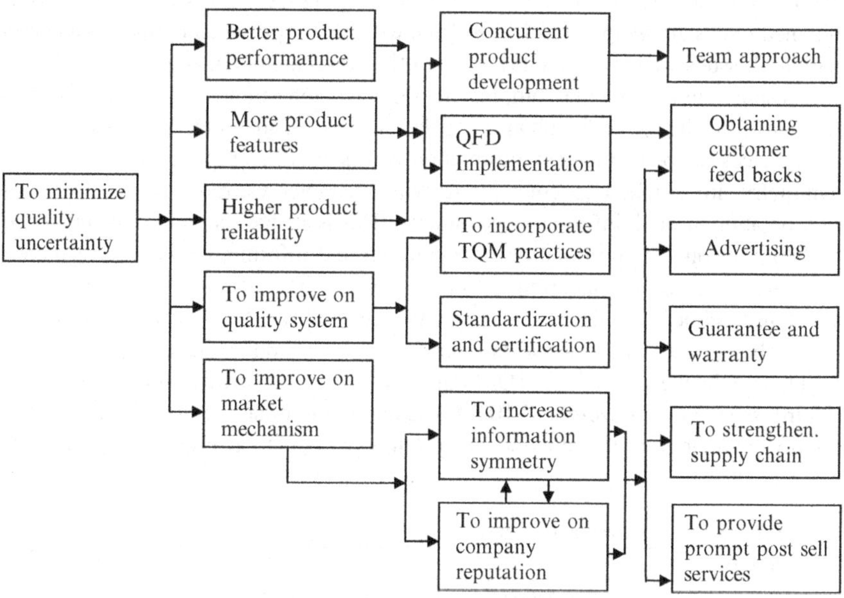

Fig. 5.10 Tree diagram to minimize quality uncertainty

factors needed for improving the market mechanism. Information asymmetry could be surmounted by enhancing the reputation of the company.

Reputation helps raise the information symmetry, as advertising, guarantee/warranty, strengthening of supply chain, improvement on post-sell services, and periodical customer feedback help increase information symmetry by enhancing the corporate reputation. The last level in the tree diagram largely represents the factors for improving the market mechanism. This level offers the bare facts for minimizing quality uncertainty. Although the list of factors for market mechanism cannot be exclusive, attempting any single factor adds to the improved market mechanism to minimize the overall quality uncertainty.

5.6.2 *Prioritization of the Issues*

Priority matrix should be developed to determine the priorities of the issues. It is a quantitative solution obtained through the analysis of ID. Similar factors can be clubbed and a priority can be determined for such combinations. Priorities are also on the basis of their cost and the ease of execution. Looking at these aspects, the priority based action plan can be drawn. Probabilities of quality uncertainty are used to determine the priorities of the implementation issues. Related but scattered issues are clubbed together for prioritization.

This clubbing of the related issues reduces the efforts required to minimize quality uncertainty. Only endogenous factors have been considered for prioritization. Socio-economic issues are not considered while developing the prioritization, whereas the retained issues are weighed at 100%. Then weights for prominent issues or a combination of issues are computed. Priorities of the issues are obtained in Table 5.14.

As shown in the tree diagram, information symmetry builds the reputation and at the same time the former is enhanced by the latter. Also, guarantee/warranty and post-sell services can be included in information symmetry. So, reputation and information symmetry are clubbed together along with guarantee/warranty to a single issue of reputation through information symmetry. This issue has the highest weight of 50.

Working on the combination of reputation and information symmetry thereby minimizes the quality uncertainty to the extent of 50%. Product development and QFD are clubbed together and customers' feedback is also incorporated in QFD. Thus, the second largest issue is coined as robust product development through QFD, with the weight of 25.7. TQM implementation and certification are clubbed as a third issue, with the weight of 10.30. The remaining issues in descending order are supply chain, logo or brand name, and product price, with a collective weight of 14.

Table 5.14 Prioritization of issues

Priority	Issues	Weight
1	Reputation through information symmetry	50
2	Robust product development through QFD	25.7
3	TQM implementation and certification	10.30
4	Supply chain	6.35
5	Logo or brand name	3.83
6	Product price	3.83
Total		100

5.7 Contributions to the Theory of Quality Uncertainty

The ID helps find the root causes of quality uncertainty while the solution guidelines are offered in the tree diagram. Prioritization of the issues then provides an action plan. The summary of the various contributions to the theory of quality uncertainty follows:

- The ID generates the data, which can be used for failure analysis of quality uncertainty.
- Quality uncertainty can be expressed as a probability of quality failure. Various techniques of reliability engineering can be applied to determine the extent of quality uncertainty.
- Low reputation of the company, nonenquiry of customers' feedback, information asymmetry, and lack of TQM practices are the prominent players involved in quality uncertainty.
- A sector-wise study reveals that the market mechanism generates nearly half of the quality uncertainty, whereas the lack of a quality system adds nearly 30%. It means "quality uncertainty and market mechanism" of Akerlof holds substantially.
- Lack of application of TQM practices and nonenquiry of customers' feedback are common drivers as well as outcomes and therefore require special attention.
- Developing nations experience the phenomenon of quality uncertainty to the extent of 35% because of the socio-economic causes only. Hence, quality uncertainty is lesser by 35% in developed nations than in the developing nations.
- Reputation through information symmetry and robust product development through QFD form 75% of the action agenda to minimize quality uncertainty.
- Quality uncertainty is observed following a distribution from the family of exponential distributions.
- A simpler method for minimizing quality uncertainty useful for an industry or a firm is developed as shown in Fig. 5.11. Focusing only on relevant aspects and assigning suitable weight to each aspect would make this method industry specific.

5.8 Managerial Implications of the Root-Cause Method

The root-cause analysis of quality uncertainty reveals facts relevant to practical implementation. Findings indicate that the market mechanism is largely involved in quality uncertainty. So, efforts made toward the success of product quality should equally be made at the market end. It suggests that a company can succeed by producing high-quality products as well as by ensuring that the customers realize the product quality. Reputation of the company is observed to be the foremost factor that minimizes quality uncertainty. Hence, every effort must be made to improve the company reputation. Interaction with customers through frequent feedbacks and

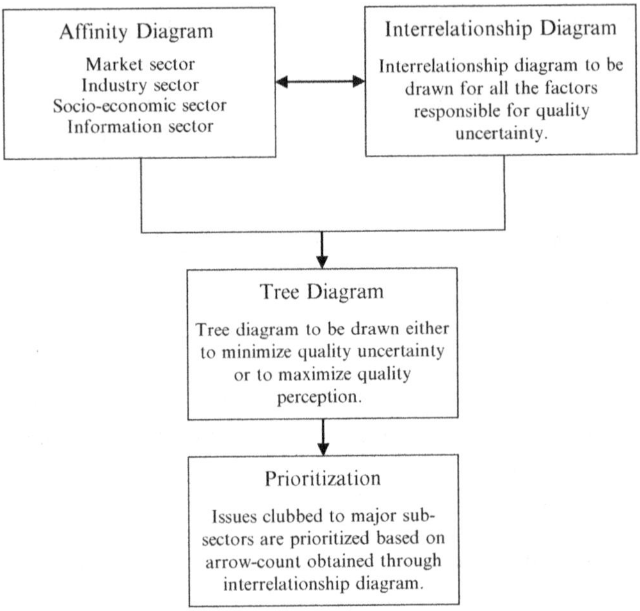

Fig. 5.11 Simpler method of minimizing quality uncertainty

subsequent product development are more practicable solutions for this. Moreover, it was established that quality uncertainty is also dependent on socio-economic conditions.

Quality uncertainty could be analyzed like other failures in reliability engineering. Quality uncertainty becomes the probability of quality failure. Here, quality uncertainty is represented by the probability tree diagram. A suitable probability distribution function from the family of exponential distributions represents the quality uncertainty behavior. It helps study quality uncertainty scientifically rather than through mere managerial speculation.

Quality uncertainty analyzed in this work is generic to developed and developing nations. Analysis of company-specific cases of quality uncertainty will depend on the prevailing conditions. For this, all aspects should be weighted systematically – on the basis of the experiences, trade volume data, and expertise. Accordingly, attempts should be made to minimize quality uncertainty through the product life cycle.

Appendix B

Worksheets 3 and 4

Using Worksheet 3

Worksheet 3 refers to the first part of the computation for root cause analysis of quality uncertainty. Worksheet 3 is an interrelationship matrix. Entries in worksheet 3 are as follows:

1. At the first place identify the issues that cause quality uncertainty of a product or products of the company. Make a list of the factors for every identified issue.
2. Make a list of all such factors identified so far. Number the factors in A, B, C, D... format. Now each factor will become a member of the interrelationship matrix in worksheet 3 in A, B, C, D... format.
3. Ponder carefully on relationships of the factors with each other. Give a thought on every possible pair of factors. There may be relationships in some pairs of the factors and for some of the pairs no relationship may exist.
4. For every pair where a relationship exists, identify cause (the factor that is responsible) and effect (the factor that is outcome). Draw arrows from cause to effect in the interrelationship matrix for all such pairs. Do this exercise only once for each possible pair. Hence, only half of the matrix, along a diagonal, is filled.
5. Using past experience, attach a suitable weight to each factor.
6. Now, there are incoming and outgoing arrows for each factor. These are seen in the row as well as the column of the factor. Count the incoming and outgoing arrows for each factor from the row and the column. Multiply the counts by the weights of the factors and make the entries – In, out, and total in the last three columns.

Using Worksheet 4

Worksheet 4 takes us toward the solution to the problem of quality uncertainty. The entries of incoming arrows are important for knowing the failures of quality uncertainty whereas outgoing arrows indicate the causes. Incoming entries can be translated into probability of quality uncertainty (P_{qu}). Entries in worksheet 4 are as follows:

1. Put the identified issues in the first column of worksheet 4.
2. List down the related factors for every issue in the second column of worksheet 4.
3. Compute P_{qu} for each factor in the third column of worksheet 4. For example P_{qu} for the factor A is as given below:
   P_{qu} (A) = Incoming arrows for A (from worksheet 3)/total incoming arrows (from bottom of worksheet 3)
4. Similarly, compute P_{qu} (issue) for each issue in the fourth column of worksheet 4 by adding P_{qu} of the all related factors to the issue.
5. Develop a solution as a priority matrix which is based on P_{qu} (issue) in worksheet 4. The issue with highest P_{qu} is a highest priority issue with a rank 1 in

priority matrix. Similarly, list up to five priorities in the priority matrix of worksheet 4.

6. Make an action plan for the priority issues to minimize the quality uncertainty.

Worksheet 3

WORKSHEET 3

	A	B	C	D	E	F	G	H	I	J	K	L	M	N	O	P	Q	R	S	T	U	V	W	X	Y	Z	In	Out	Total
A																													
B																													
C																													
D																													
E																													
F																													
G																													
H																													
I																													
J																													
K																													
L																													
M																													
N																													
O																													
P																													
Q																													
R																													
S																													
T																													
U																													
V																													
W																													
X																													
Y																													
Z																													
Total																													

Worksheet 4

WORKSHEET 4

Issues	Factors	P_{qu}	P_{qu} (issue)

P_{qu}

Priority	Issues	Weight (P_{qu})
1		
2		
3		
4		
5		

Chapter 6
Dynamics of Quality Perception

The term quality uncertainty suggests that TQM may succeed or fail, totally or partially. Level of TQM implementation determines the level of quality uncertainty. If quality failure is quality uncertainty, then what is quality success? The term becomes quality perception. The way customers perceive product quality is quality perception. Quality perception is a systems output when total quality management index (TQMI) and information symmetry are system inputs.

System dynamics modeling of quality perception is introduced in this chapter. System dynamics encompasses all areas of study and proved to be highly satisfactory in resolving many national and international issues. The issue of quality in various perspectives is examined from the early stage of system dynamics (Jermain 1963). Sice et al. (2000) analyzed duopoly competition for quality. Khanna et al. (2004) applied system dynamics to the TQM transition journey of the automobile sector in India. Vojtko et al. (2006) have demonstrated a preliminary structure on market behavior. System dynamics is a suitable technique for representation and modeling of business processes.

A new framework for quality perception is provided in this chapter. Simulation runs are carried out for the general framework by using the available data in India. Conclusions are drawn for important parameters. The effects of market side enablers on quality perception are analyzed. General framework is then elaborated to the major sectors of quality perception. Dynamic hypothesis and boundary selection for real life scenario are presented. Also, required model structure is prepared, and simulation runs are attempted. Perception patterns have been generated for the changing level of information symmetry for developed and developing nations. The impact of advertising and reputation on quality perception has been observed by simulating the sector wise models.

The process of quality perception is explained through simulations. The impact of various parameters on quality perception is shown by obtaining the graphs through simulations. The impact of economic condition and literacy level has been assessed in the case of developing nations. This theoretical foundation provides an insight for increasing quality perception for improved business performance.

L. Wankhade and B. Dabade, *Quality Uncertainty and Perception*,
Contributions to Management Science, DOI 10.1007/978-3-7908-2195-6_6,

6.1 System Dynamics Methodology

The system dynamics methodology helps crystallize the complex problem under study and generate the scenario that is similar to a real-life problem. By changing dependent values, one can see into the future spectrum of a particular event happening. Also, the event's repercussions on all other parameters can be studied. Succinctly, system dynamics is a method for prognosis of an issue under consideration in all possible perspectives.

Forrester from the Massachusetts Institute of Technology is the founder of system dynamics. The starting of system dynamics could be traced to Forrester's (1961) book titled Industrial Dynamics. He blended the traditional management and feedback control of electrical engineering with the modern technique of computer simulation. Industrial Dynamics was widely accepted as well as criticized. Forrester came out with more publications on systems approach. Forrester also wrote Principles of Systems (1968), Urban Dynamics (1969), and World Dynamics (1971). Soon, system dynamics became a popular term for systems approach and modeling of a complex world. System dynamics developed into a full-fledged discipline all over the world, through the 1980s and 1990s.

The system dynamics methodology starts with system thinking by using causal loop diagrams and polarity-based feedback. These mapping techniques convey a pictorial representation of the problem. The modeling task is carried out by means of structured stocks and flows, and detailing of the variables. Flows, levels, rates, constants, and auxiliaries are used in system dynamics modeling. Integration of flow at some rate results in the accumulation of level, which is the basic philosophy underlying system dynamics. Feedback loops, delays, and multiplicity of relationships make the model live and dynamic. The model works for real-life data through simulation runs. System dynamics modeling is a creative process. Various softwares like Vensim, mystrategy, Powersim, Stella, and Ithink are available for the system dynamics modeling.

6.1.1 Procedural Steps in System Dynamics Modeling

"There is no cookbook recipe for successful modeling; no procedure you can follow to guarantee a useful model," said Sterman (2000) while elaborating procedural steps. But, some guidelines are required to have a disciplined approach. The procedural steps are as follows:

- The modeling starts with problem definition. We make clear the purpose of the modeling task on the basis of the gravity of the problem. The goals should be fixed so that modeling and simulation should not become a fishing expedition.
- Once the problem is defined, we need to enter the articulation process. Elaborate on the theme and identify the key variables. Also, rough behaviors of the key variables may be drawn to understand the required timeframe for the modeling.

- It's a time to work on the dynamics of the problem. Identify the endogenous, exogenous, and excluded variables. Draw a causal loop diagram which is then translated into a stock and flow model. Constants and variables are embedded by using simple logics and arithmetic. Feedback loops, delays, and multiplicity of relationships make the model live and dynamic.
- Now, it's time to evaluate the model. If it is felt that the model is not producing reasonable behaviors, then examine each relationship and equation carefully. Conventional tests can be performed to check the robustness of the model.
- Now, it's the time to improve the model. Make a few changes, as desired, on the limiting variable or resource to observe the changed performance. Changes should be made according to the preferred performance. The insights gained from simulations are further useful to reiterate the model toward perfection.

6.2 Causal Loop Mapping of Quality Perception

6.2.1 What is a Causal Loop Diagram?

A causal loop diagram is a system mapping technique. It exhibits the system elements and their relationships. All the elements from a system are drawn as per their attachment to each other. An outwardly arrow is drawn from an element to another one which indicates that some relationship exists between these two elements. An element with an emerging arrow is an influencing variable. It influences the element that receives the arrow. Thus, the relationships are shown for all possible pairs. An element may have multifarious relations in a causal loop diagram.

Further, "+" or "−" is used to show the type of influence in the pair. A positive sign means that a change in the variable effects a similar change on the influenced variable, whereas a negative sign indicates an adverse impact on the influenced variable. Thus, a causal loop diagram represents a system which anybody can read with a sufficient level of understanding.

6.2.2 Causal Loop Diagram of Quality Uncertainty or Perception

Figure 6.1 shows the causal loop diagram of quality uncertainty. Major sectors are quality management, socio-economic, and marketing sectors. Information asymmetry, shown centrally, is the main element of the system. Similarly, another important element of the system is quality uncertainty due to information asymmetry. As shown, the factors on the left are the socio-economic factors that generate the general part of the information asymmetry, whereas the factors above information asymmetry are market enablers responsible for product part of the information asymmetry. The behavior of high quality product or plum is a main object of study, which is represented as "Market for Plum" in this causal loop diagram.

Fig. 6.1 Causal loop diagram of quality uncertainty

Quality uncertainty or perception of a plum, as shown in the causal loop, is firstly determined by the amount of TQM implementation that makes the product quality, and the amount of information asymmetry. The level of TQM implementation has been directly responsible for the quality that is being manufactured. Higher level of TQM implementation will reduce the quality uncertainty at the first level. The TQM implementation depends on the TQM constructs that are shown on the right side of the figure. All the TQM constructs improved together produce a high quality product and hence probability of acceptance of this product in a market becomes higher. This minimizes quality uncertainty at the place of origin.

Subsequently, once the product has been received in the market, the force of information asymmetry plays a role in quality uncertainty or perception of the product. Here is a need to raise the level of product information provided to the customers so that information asymmetry will be lowered to reduce the quality uncertainty at the second place. Various marketing strategies or parameters are helpful to increase product information symmetry. These are shown above information asymmetry in the causal loop diagram. If quality uncertainty of a plum is reduced, the market for the plum increases which ultimately restricts the market for lemons.

Information asymmetry is generated by two sets of factors – socio-economic and marketing. The factors from socio-economic set are largely responsible for information asymmetry in developing nations. And similarly, marketing strategies play an equal role to control information asymmetry. Hence in developing nations presence of information asymmetry becomes an obvious fact and has inherent limitation in minimizing information asymmetry and quality uncertainty.

Unlike developing nations, information asymmetry in developed nations has been largely dependent on the dissipation of product information and hence can be controlled completely by using appropriate marketing strategies. Therefore, quality uncertainty of high quality products in developed nations can be ascribed to limitations in the amount of TQM implementation and in the use of marketing enablers by the companies.

6.3 Preliminary Model, Variables, Data, and Analysis

Quality perception is a complex phenomenon involving social, cultural, economic, and technical aspects. System dynamics is a suitable technique for modeling and analysis of quality perception. Figure 6.2 is a general model of the TQM with quality perception; this model is programmed in Vensim PLE. The model is preliminary in content but quite sufficient to investigate into the important managerial aspects. Polarities in the model indicate the nature of effect. The arrows depict the direction of effect. The positive signs indicate similar kind of change from one variable to another, whereas the negative signs indicate contradiction. The model is meant for a representative player with an average TQMI. Important variables are briefly explained.

6.3.1 *Information Symmetry*

Information symmetry is assumed on the basis of the level of education and information proliferation. Population growth may curb the information flow to a little extent. State enactments have to be used to effect the information flow. The cultural milieu may produce a positive or negative impact, which varies from society to society.

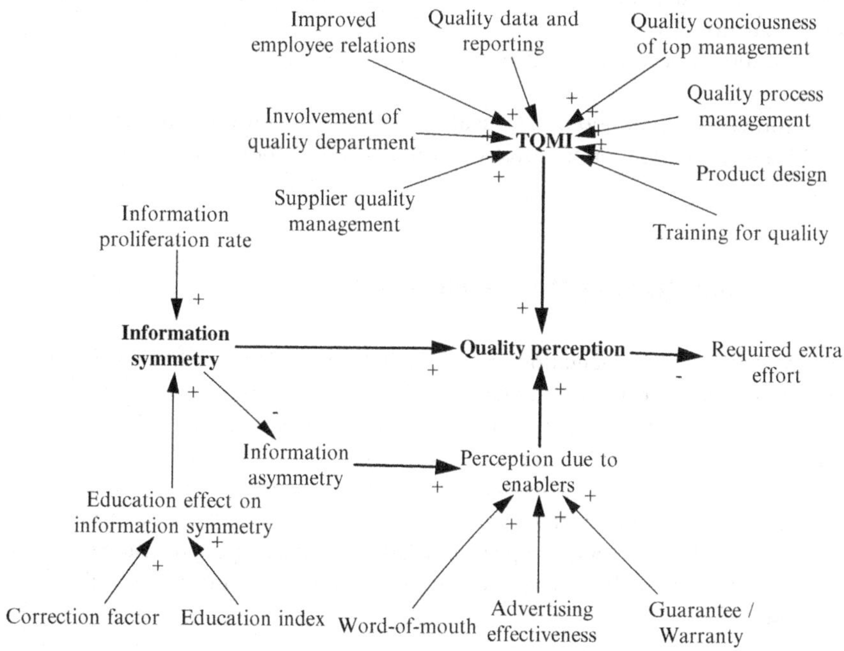

Fig. 6.2 System dynamics model of quality perception

The education level is never translated completely into information symmetry; many other factors play a sizable role. Nevertheless, education is a prominent determinant exerting its influence through a correction factor and resulting in "education effect on information symmetry." Quantification of many other causal factors is a complex issue. Information proliferation primarily depends on word-of-mouth, advertising, and media effect. However, these are termed enablers for considering the means of information proliferation more immediate to market side. The information proliferation rate becomes a function of the cultural milieu, population growth effect, national economy, state enactments, technological advancements, and the working of supply chain because word-of-mouth and advertising are taken to the market side as enablers of quality perception. Asymmetric customers (lacking in information or education) have a chance exposure to information. Gathering knowledge through the information proliferation makes a customer symmetric to product quality. Hence, the rate at which information proliferates determines the information symmetry. Thus, the information symmetry equation is arrived at in following steps.

Information proliferation rate $= f$ (advertising, word-of-mouth, cultural milieu, population growth effect, national economy, state enactments, technological advancements, and working of supply chain).

Using advertising and word-of-mouth as enablers, the information proliferation rate becomes as follows:

Information proliferation rate $= f$ (cultural milieu, population growth effect, national economy, state enactments, technological advancements, and working of supply chain).

Education effect on information symmetry = Education index $\times$ Correction factor.

Information symmetry = Education effect on information symmetry + (1 − Education effect on information symmetry) $\times$ Information proliferation rate.

6.3.2 *Total Quality Management Index*

The percentage implementation of TQM in industry or TQMI primarily determines the success rate in establishing the product or profit shares in the market. TQMI is an important component along with several other critical factors (Motwani 2001). According to Badri and Davis (1995), "Quality success of any industry or company can comprehensively be studied on the basis of critical evaluation of major factors." Each factor is assessed separately. Table 6.1 presents a list of these factors. TQMI is a mean of all these factors (Joseph et al. 1999). It is considered a measure of product quality. Product quality in any other measurable term and value suffices the purpose.

$$\text{TQMI} = \sum \text{TQM factors}/8.$$

Table 6.1 Factors for quality management and perception

Quality management	Quality perception
Role of top management	Information symmetry
Role of quality department	Education
Training for quality	National economy
Product/service design	Cultural milieu
Supplier quality management	Global/national enactments
Process management/operating procedures	Advertising
Quality data and reporting	Guarantee/warranty
Improved employee relations	Word-of-mouth
	Company's reputation
	Working of supply chain

6.3.3 Perception Due to Enablers

Enablers perform the market-end role of enhancing quality perception by counteracting prevailing asymmetry. Asymmetric customers are translated commensurately into symmetric ones by the mixed use of enablers. Advertising, word-of-mouth, guarantee, and warranty are the enablers used to enhance the quality perception.

$$\text{Perception due to enablers} = \text{Information asymmetry} \times (\text{Advertising} + \text{Word-of-mouth} + \text{Guarantee/Warranty}).$$

6.3.4 Quality Perception

Quality perception is a main component of this model. TQMI, information symmetry, information asymmetry, and perception due to enablers are the inputs to quality perception. TQMI and information symmetry are directly responsible for quality perception. However, quality passes through prevailing information, and results in quality perception and quality uncertainty. Quality lost in transmission from industry to market, due to asymmetry phenomenon, is a quality uncertainty. Enablers are used to make up for this quality loss. The feasible reversion or adjustment of quality loss, by the use of enablers, leads to perception due to enablers. Hence, quality perception becomes a combined function of information symmetry, TQMI, and perception due to enablers.

$$\text{Quality perception} = (\text{TQMI} \times \text{Information symmetry}) + \text{Perceptiondue to enablers}.$$

6.3.5 *Required Extra Effort*

Enablers enhance the quality perception. But enablers may not be sufficient for 100% quality perception, and perception loss may still be noticed because of asymmetry phenomenon. "Required extra effort" of this model is the opposite of quality perception.

$$\text{Required extra effort} = 1 - \text{Quality perception}$$

Some extra tools are still necessary to make up for the quality perception loss. Dynamic customer relationship management, defined in informational parlance by Park and Kim (2003), target marketing or database marketing, QFD, and price signaling are the identified tools currently available for this purpose. Traditional QFD focused on operational issues. But the efficacy of traditional QFD in taking marketing horizon to manufacturing parlance was a limited one. Now, the newly coined term "QFD strategy house" (Gonzalez et al. 2004) translates marketing strategies into manufacturing strategies to improve upon the quality perception loss. Although QFD tries to build in quality perception at the product development stage, price signaling is a policy decision at the launch of the product in the market. Target marketing or database marketing caters finally to an extra effort to marketing.

6.3.6 *Purpose and Data Collection*

The purpose of this work is to attempt the formulation of general information symmetry, to relate TQM status, to arrive at quality perception, and to observe market-side repercussions on quality perception. The modeling aims to provide a system dynamics framework for quality perception and to observe the role of the changing level of market-side enablers on quality perception. Inputs to this model are current facts of TQM constructs, the education index, and the rate of information proliferation. The values of market-side enablers are tested, and their impact on quality perception along with efforts required is investigated. The combined values of enablers taken for experimentation are 00, 30, 60, and 90%. Variable values are taken by judgment, where quantification is not feasible. Simulation runs are carried out with some data testing.

India is considered a specific case of developing nations. Jagadeesh (1999) surveyed and published Indian TQM factors for the year 1996 (Table 6.2). For the purpose of modeling, factors are translated into a 0–1 scale, and extrapolated to the year 2000. Khanna et al. (2004) concluded that for a moderate to weak scenario, 13 years would be required to attain TQM maturity in the Indian automobile sector. We assume a varying maturity period of 15–20 years for a general case of Indian TQM factors. The maturity value is expressed here to the extent of 90–92%. This is

Table 6.2 Indian TQM factors with values and perceived maturity period

TQM factors	Value (1996)	Maturity (years)
Role of top management	0.314	19
Role of quality department	0.372	18
Training for quality	0.262	20
Product/service design	0.418	17
Supplier quality management	0.428	16
Process management/operating procedures	0.452	15
Quality data and reporting	0.416	17
Improved employee relations	0.416	17

due to the fact that 100% achievement is seldom feasible. As shown in Table 6.2, maturity period variations are due to the base-level variations in values of TQM factors.

The literacy value for the year 2000 is taken from the Indian census report. The model assumes that literacy is increasing at a constant rate, on the basis of the previous data computation. The literacy is translated into information symmetry through a correction factor.

6.3.7 *Analysis from Simulation Runs*

Analysis is performed for key variables such as TQMI, information asymmetry, perception due to enablers, quality perception, and required extra effort. The graphs taken for analysis are direct output of the simulation runs. Inbuilt explanatory notes appear under each graph. Figure 6.3 shows the behaviors of information symmetry, asymmetry, and TQMI. In 2000, information symmetry obtained is 0.5212 and increases to 0.814146 (slope of 1.2% per year) by 2025, whereas asymmetry is higher to the extent of 0.4788 in 2000 and declines to 0.1858 (slope of 1.2% per year) by 2025. The behaviors of symmetry and asymmetry, on a larger time scale, are exponential growth and decay, respectively. Both symmetry and asymmetry behave linearly for a shorter time span of this model. As a goal setting, TQMI displays a straight line (slope of 3% per year), saturating to a maturity value by 2020.

For various values of enablers, perception due to enablers is sufficiently high at the simulation start, but declines by 2025 (Fig. 6.4). For a lower value of enabler, the descending rate is lower; for higher enablers, the descending rate is higher. For a 30% enabler, perception due to the enabler is 0.1436, declining to 0.0557 with a slope of 0.003526 per year; for a 60% enabler, it is 0.2872, declining to 0.1115 with a slope of 0.007068 per year and for a 90% enabler, it is 0.4309, declining to 0.1672 with a slope of 0.010548 per year. This trend is observed because of the effect of information symmetry on perception due to enablers, signifying that for lower values of information symmetry, more enablers are needed to enhance the quality

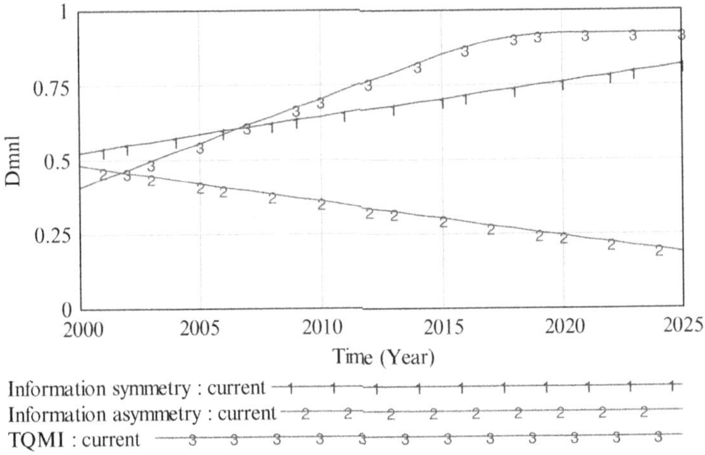

Fig. 6.3 Symmetry, asymmetry, and TQMI

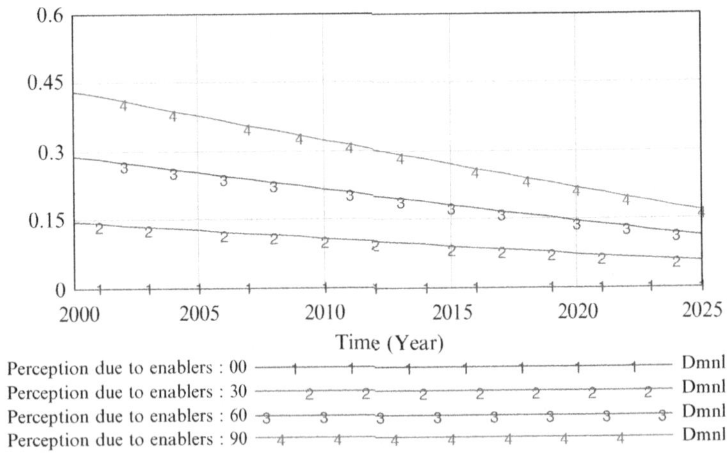

Fig. 6.4 Perceptions due to enablers

perception and vice versa. Rise in information symmetry breeds the quality perception and, accordingly, the role of enablers diminishes.

"Perception due to enablers" is an important vehicle in counteracting asymmetry phenomenon. Figure 6.5 clearly shows direct relationship between perception due to enablers and information asymmetry. As asymmetry increases, enablers have greater roles to contribute toward "perception due to enablers." For higher values of information asymmetry, higher values of enablers are required to speed up the process of perception. This is explicitly noticed from the slopes (Fig. 6.5). For a 30% enabler, slope observed is 0.30; for a 60% enabler, slope is 0.60; and for a 90%

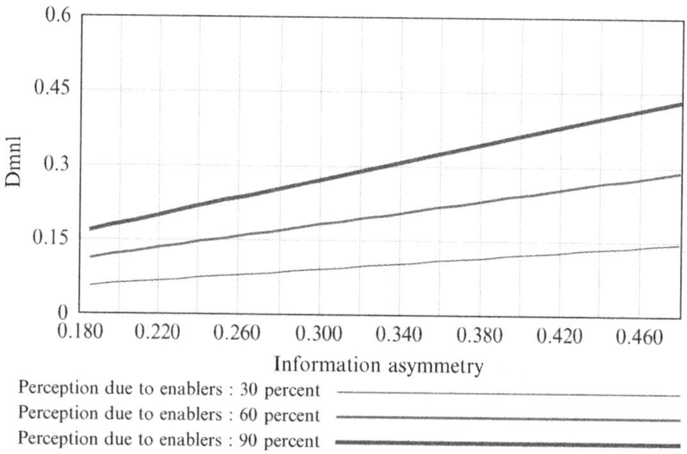

Fig. 6.5 Information asymmetry and perceptions due to enablers

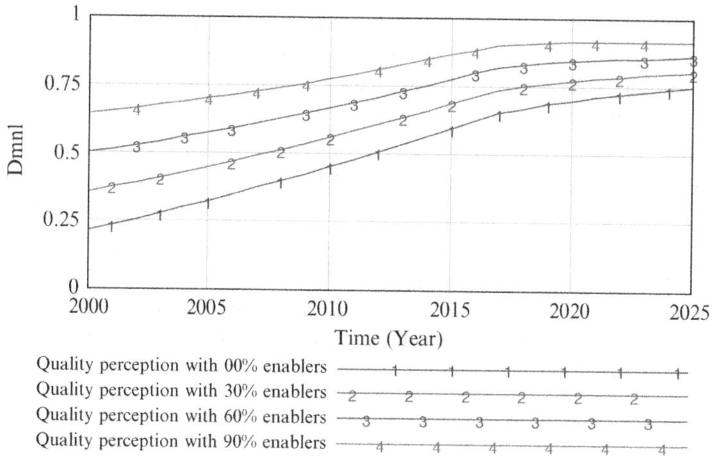

Fig. 6.6 Quality perceptions with varying enablers

enabler slope is 0.90. To make the product quality perceivable by the large section of customers, the management should decide on the deployment of enablers' mix, on the basis of the established relationship between information asymmetry and perception due to enablers.

Perception due to enablers coalesces to the probabilistic formulation of TQMI and information symmetry. The pattern obtained of quality perception (Fig. 6.6) is a typical goal-seeking type of system dynamics. The goal to be achieved is 100% (i.e., 1 on a 0–1 scale). The quality perception has been smoothly saturating nearer to an achievable goal in the zone of 2015–2020. Quality perception follows the combined path of symmetry and TQMI. As seen in Fig. 6.6, the pattern indicates the

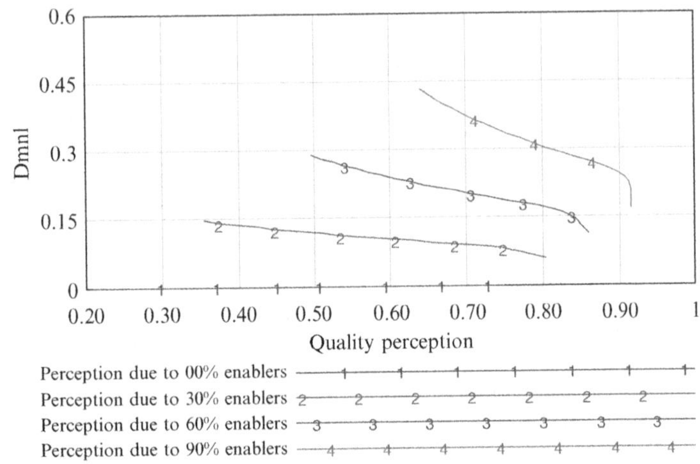

Fig. 6.7 Quality perception and perceptions due to enablers

predominance of saturating TQMI in quality perception. Also, it is observed that information symmetry and enablers enhance the quality perception significantly. As soon as simulation time increases, quality perception saturates along with the rising information symmetry.

In Fig. 6.7, perception due to enablers is plotted against the quality perception, where the extent of importance of perception due to enablers is made explicit. Perception due to enablers plays a vital role at lower values of quality perception. Toward higher quality perception, "perception due to enablers" starts diminishing as "perception due to symmetry" takes over.

The last component of this experimentation is to find out the amount of extra effort required to make up for the quality perception loss. "Required extra effort," drawn against the backdrop of the changing dynamics of information symmetry, shows a graph in Fig. 6.8. Rising symmetry gives way to decay in the efforts required. Hence, "required extra effort" largely depends on the percentage implementation of enablers and declines in a smooth way. So enablers are quite helpful and sufficient to make up for the perception loss. However, the graph also suggests the fact that at the saturating end it is not wise to invest more on enablers.

6.3.8 Findings from the Simulation Runs

System dynamics model developed here helps in behavioral study and analysis of quality perception. The conclusions are the following:

- Both TQMI and information symmetry are crucial in determining the quality perception. TQMI and information symmetry make the probability of quality perception.

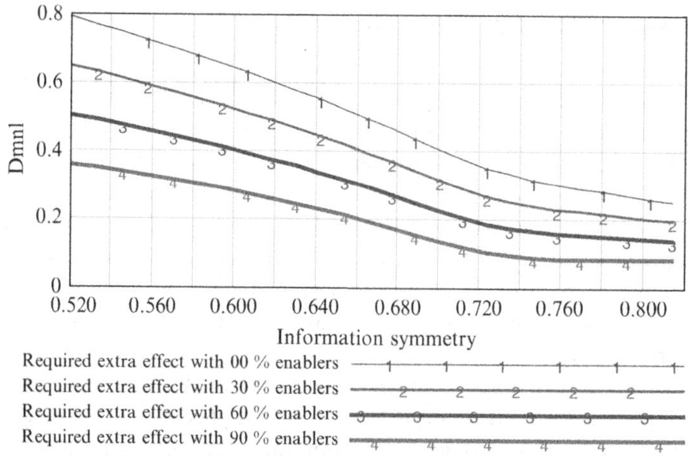

Fig. 6.8 Required extra efforts along with enablers

- Enablers such as advertising, word-of-mouth, guarantee, warranty, and rebate play an important role in enhancing quality perception.
- Rise in information symmetry breeds the quality perception and accordingly, the role of enablers diminishes. "Perception due to enablers" starts diminishing as "perception due to symmetry" takes over.
- The use of enablers converts asymmetric customers commensurately. Target marketing or database marketing should be used to further translate asymmetric customers into symmetric ones.
- Total quality perception is expressed as follows:

$$\begin{aligned}\text{Total quality perception} = {} & \text{Perception}_{\text{symmetry}} + \text{Perception}_{\text{advertising}} \\ & + \text{Perception}_{\text{word-of-mouth}} + \text{Perception}_{\text{guarantee/warranty}} \\ & + \text{Perception}_{\text{rebate}} + \text{Perception}_{\dots\dots\dots} \\ & \leqslant 1 \text{ (i.e., 100\% of TQMI or 100\%} \\ & \text{of product quality).}\end{aligned}$$

- The total quality perception equation implies that enablers could be used only to the extent of 100% or optimal quality perception. Still increased use may not help in the further enhancement of quality perception, but may affect profit margin and ultimately result in revenue loss to the company. Hence, attempts should be made to maximize quality perception by maximizing TQM efforts and optimizing enablers in concurrence with information symmetry.

6.4 Toward Market Dynamics

Products, supply chain processes, and customers are major market components. The flow of information largely determines the decision processes. Recently markets are more pronounced in the sense that many subtle and traditional market parameters like product deliveries, customer feedbacks, customer complaints, product choices, etc., are incorporated in some or other modern parameter. For example, TQM is a wider concept where many customer aspects are taken into consideration. TQMI or representation of product quality is a crucial input for modeling the market scenario. Thus, prevailing information symmetry and quality perception are important factors for market study.

Market is really a huge complex process. It is dynamic with multifarious relationships. Quality perception in itself is a complex part of market dynamics. But for being so, we should not desist from modeling of quality perception. The foundation of market dynamics can be laid on product quality, information flow, and quality perception. Gradually, more factors and relationships could be incorporated to make it dynamic in real sense.

Market, information symmetry, company reputation, and socio-economic milieu are the major identified sectors for modeling. Socio-economic submodel pertains to developing nations. Most of its parametric relationships and data are available in the literature on development economics. Research work on modeling of market scenario is meager in content, but basic theories are helpful. Yet, most of the market relationships are to be crystallized. Major challenges lie in modeling the sectors of information, company reputation, and their linkages to market behavior.

6.4.1 Information Sector

Information symmetry and resulting quality perception constitute a complex phenomenon. This submodel can be started with adaptation to Bass diffusion model (Bass 1969; Bass et al. 1994; Sterman 2000). It includes the basics of information proliferation called adoption rate of the product. Adoption from advertising and adoption from word-of-mouth are major components of Bass diffusion model which becomes the starting of the information sector. Information symmetry is more important than the product diffusion. Other than adoption from advertising and from word-of-mouth, the effects of guarantee, warranty, and rebate should be realized. These effects do not contribute directly to information symmetry. Separate submodels should be developed to translate their effect into information symmetry.

Even advertising effect cannot be simpler than the effect of Bass diffusion model. Required repetitive exposure, advertising quantum, changing structure of media, and many other variables (Tellis 1988; Tellis and Fornell 1988; Tellis et al. 2005; Iyer et al. 2005) determine the advertising effectiveness. Sometimes a firm may float the quality or price information differently, but its effectiveness may be limited (Moorman 1998, 2005). Also information symmetry depends on consumers' learning abilities and their decision processes in buying behavior (Tellis and Prabhu 2000). It is a customers' side which must predominate while modeling information symmetry.

6.4.2 Company Reputation

A range of issues are required for modeling of company reputation. Primarily reputation depends on market share, product quality (extent of TQM implementation), product price, post-sell services, customer feedback, and corporate policies – including social and welfare activities. Advertising, media reports, and word-of-mouth are instruments that enable information symmetry to enhance the company reputation. Market share is zero for an exclusively new entrant. Data available in consumer reports can be used for modeling of reputation. All other competitors in the same product category are involved in market dynamics. Product price is also one of the observed factors in a competition (Tellis 1989). The works of Dybvig and Spatt (1983), Shapiro (1982, 1983), Fudenberg and Kreps (1987), Fudenberg and Levine (1992), Fombrun (1996), and Horner (1999) are helpful for modeling of reputation. Wide array of reputation-related factors and ample literature on each factor are available. However, independent modeling approach seems suitable for the modeling of reputation. The literature should be referred to for obtaining the guidelines.

6.5 Quality Perception and Market Dynamics

6.5.1 Dynamic Hypothesis and Boundary Selection

Information symmetry is an identified key factor responsible for quality perception. Product quality at market entry is asymmetric to the customers if product is new to the company and becomes totally asymmetric if the company itself is a new entrant. Improvement of product quality over the years builds up the reputation. Among various forces, advertising and word-of-mouth are important for information flow. Here, it is necessary to decipher the dynamics of advertising, word-of-mouth, and inherent changes in product quality (Mitra and Golder 2006) and their reflection on sales. Information asymmetry is prevalent in developing nations. Population growth, illiteracy, developing economy, lower purchasing power parity, cultural conflicts, etc., add gravity. Presence or absence of TQM culture in the organization is also an issue of concern.

Product life cycle summarily determines the extent of model timeframe. Product life is experiencing sharp decline with huge competition in the market. Time horizon of the model depends upon product type and its life cycle. Three or four years' time span is quite sufficient to carry out policy experimentation and requisite corroboration. The model can be general or for a specific industrial establishment. In a sequel to modeling process, boundary chart (Sterman 2000) is prepared as in Table 6.3. Focused parameters immediate to the model fabric are labeled endogenous, with semblance of outward disposition are exogenous, and seemingly feeble are excluded.

6.5.2 Modeling Framework

Figure 6.9 shows the generic framework for modeling. Framework has three major components, viz market, industry, and information proliferation. Reputation is a part of the module of industry/company. But, reputation is treated as a separate

Table 6.3 Boundary chart

Endogenous	Exogenous	Excluded
Product quality (TQMI)	Information symmetry or asymmetry	Population growth
Company reputation	Market share	Cultural effect
Brand or logo name	Word-of-mouth	National economy
Product price	Supply chain effect	State enactment
Advertising	Quality uncertainty or perception	
Guarantee/warranty	Education	
Post-sell services	Purchasing power parity	

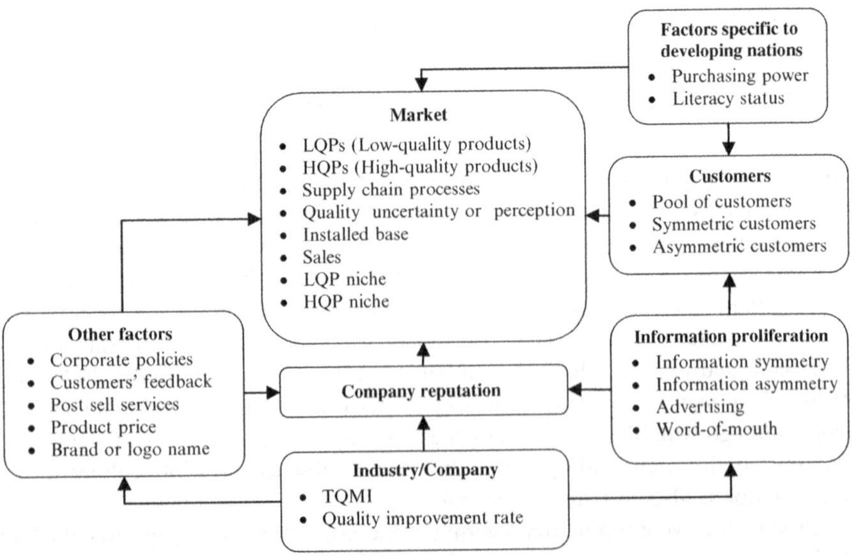

Fig. 6.9 Model structure

module for making the linkages unambiguous. Industry is a source of origin of product quality. The TQMI along with zest for quality improvement molds the product quality. Accordingly, product type in the market has broader connotation of Low Quality Products (LQPs) and High Quality Products (HQPs). Similarly, sales, installed base, and market niche are terms connected to LQP or HQP. The term "market niche" is synonymously used for market share in modeling and simulations. Important component at this juncture is information proliferation which includes symmetry or asymmetry of information and means of information proliferation like advertising and word-of-mouth. Quality passing through the process of information proliferation reaches customers and becomes quality perception at market-end.

Company reputation is a core of modeling structure. It is surrounded by three major sectors. Company's effort to enhance the product quality forms a foremost sector. Means of product-quality-information dissipation makes the information proliferation sector. Other factors like brand, price, etc., make the third important sector of company reputation. These sectors also play a direct role in market dynamics. Company reputation along with all other sectors leads to market dynamics.

In system dynamics literature, it is always said that a model should be developed at first instance. System dynamics model never becomes a complete one. Modeling, as a part of learning process, is iterative. Experiments conducted in the virtual world inform the design and execution of experiments in the real world; experience in the real world then leads to changes and improvements in the virtual world and in participants' mental models (Sterman 2000). The aim of simulating this preliminary model is to contribute to the theory of quality perception. We also, initiate market dynamics by experimenting in the virtual world. Simulations will provide an insight toward the expansion of the model. Preliminary model can be drawn in four sectors, viz market, industry, socio-economic factors, and information proliferation. Model is programmed in Vensim PLE.

6.5.3 *Focus of the Experimentation*

Like all good research, studies that develop theory through simulation should begin with an intriguing research question that reflects deep understanding of the extant literature and relates to the substantial theoretical issue (Weick 1989). Without such a question, simulation research simply becomes a "fishing expedition" in which the researcher lacks focus and theoretical relevance, and risks become overwhelmed by computational complexity (Davis et al. 2007). Literature review and quality perception exploration indicate that market dynamics model should be a huge one, with bigger submodels, which represents the entire system. It is not possible to attempt all these details in the preliminary stage of modeling. Hence, this modeling work is confined to fewer relationships. Quality perception is new to management people. The aim of modeling and simulation is to study behavior and build theory. Attempt is made to understand some parametric relationships. We are exploring the role of advertising and reputation in causing information symmetry and quality perception. Experiments are planned with a focus on information symmetry, subsequent quality perception, and market niche of HQP. The HQP position in the backdrop of information symmetry and quality perception is a theme of modeling and simulations. Thus, the theory of quality perception is developed as a part of market dynamics through the following processes:

- Behavioral study of quality perception in developed and developing nations
- Study of HQP perception due to information symmetry in market dynamics
- Accessing the impact of company reputation and advertising on market parameters

LQP and HQP are taken as representative cases of product quality in market dynamics. Company reputation at market entry, advertising, and word-of-mouth are used as major variables. Product adoption by customers depends on the dissipation of product quality information. This is achieved through word-of-mouth and advertising. As word-of-mouth depends upon the sales, adoption through word-of-mouth is generated in the model. Varying levels of advertising and company reputation are the model inputs for simulation runs. The equations in this work are simple and on the basis of the general theories of economics and marketing. Quality perception equations are on the basis of the mathematical modeling.

6.5.4 *Experimentation Plan for Simulation Runs*

Table 6.4 presents the experimentation plan for simulation runs. Advertising target factor denotes number of customers to be targeted (made symmetric) per month. Current trends of database marketing and dynamic customer relationship management (Park and Kim 2003) support this mode of advertising. Normal range of monthly advertising target used for experimentation is shown in Table 6.4. Expenditure toward marketing and sales promotion efforts can also be used instead of advertising target. Company reputation is zero for a new entrant company. The best possible reputation is 100% that is 1 on 0–1 scale. All possible reputation falls in the range of 0–1. Zero, low, and moderate reputations are used for this experimentation. The values of 0, 0.3, and 0.6 are taken for zero, low, and moderate reputation in Table 6.4. Three advertising and three reputation modes make the nine combinations. These combinations are named as A1, A2, A3, B1, B2, B3, C1, C2, and C3 for developed nations. Similarly D1 to F3 bracket combinations are for developing nations. Model assumes that market niche is divided equally between LQP and HQP at the start of any simulation run.

Table 6.4 Experimentation plan

Advertising	Reputation		
	0	0.3	0.6
15,000/month	A1 (D1)	A2	A3
30,000/month	B1	B2 (E2)	B3
45,000/month	C1	C2	C3 (F3)

6.5.5 *Market Sector*

Market sector (Fig. 6.10) is on the basis of the demand, which is assumed to be constant at 10,000 (any sizable value suffices for the purpose) units per month, and sales are distributed between LQP and HQP. Respective sales are assumed proportional to quality perception and represented by the term "market share." Installed base is an accumulation of the sales.

$$\begin{aligned} \text{LQP sales} &= \text{Demand} \times \text{LQP share} \\ \text{HQP sales} &= \text{Demand} \times \text{HQP share} \\ \text{LQP niche} &= \text{Installed base LQP/total base} \\ \text{HQP niche} &= \text{Installed base HQP/total base} \end{aligned}$$

6.5.6 *Information Proliferation Sector*

Information proliferation sector is shown in Fig. 6.11. Company reputation gets translated into information symmetry through a conversion factor. It is assumed to be one here. Asymmetric customers are of a variable level, decreasing at the rate of

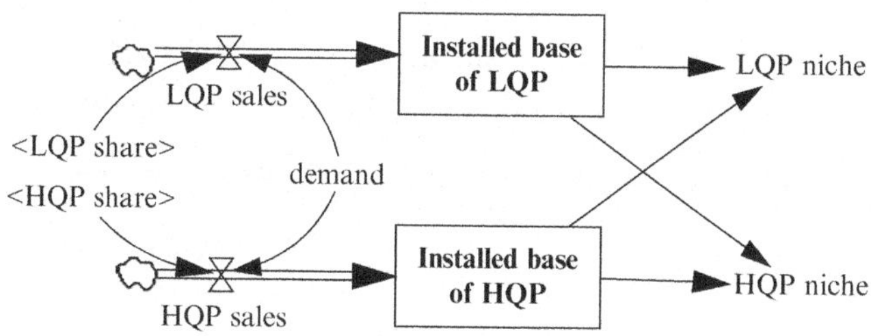

Fig. 6.10 Market sector

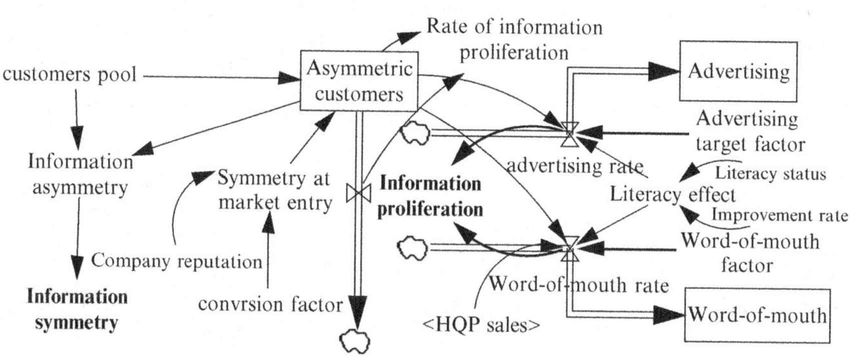

Fig. 6.11 Information proliferation sector

information proliferation. Customers' pool is a statistically sufficient value of ten million customers:

$$\text{Asymmetric customers} = \text{Customers pool} \times (1 - \text{Symmetry at market entry}) - dt \times \text{Information proliferation}.$$

$$\text{Information asymmetry} = \text{Asymmetric customers}/\text{Customers pool}.$$

$$\text{Information symmetry} = 1 - \text{information asymmetry}.$$

Information proliferation consists of advertising and word-of-mouth as level variables with advertising and word-of-mouth rates, respectively. Advertising in turn is determined by advertising target factor. Adoption from word-of-mouth is determined by HQP sales and word-of-mouth factor. On an average a customer who buys the product conveys the quality information to two customers. Both advertising and word-of-mouth go along with literacy (assumed one for developed nations) and campaign till asymmetric customers are fully converted. Component of information proliferation is negligible or conspicuous by its absence for LQPs, as the expenditure of this overshadows the low profit margin. This is quite in tune with the theory of information economics:

$$\text{Advertising rate} = \text{IF}(\text{asymmetric customers} > 0,\ \text{Advertising target factor} \times \text{Literacy}, 0).$$

$$\text{Word-of-mouth rate} = \text{IF}\ (\text{asymmetric customers} > 0,\ \text{HQP sales} \times \text{Word-of-mouthfactor} \times \text{Literacy},\ 0).$$

$$\text{Information proliferation} = \text{Advertising rate} + \text{Word-of-mouth rate}.$$

6.5.7 *Industry Sector*

Figure 6.12 is an industry sector. Quality status of HQP and LQP and their improvement rates act as inputs to this sector. Representative cases of HQPs and LQPs are taken at 0.8 and 0.4 and improve at the rate of 0.00322 to reach 0.9 and 0.5 by the maturity of the product which is roughly half the product life cycle.

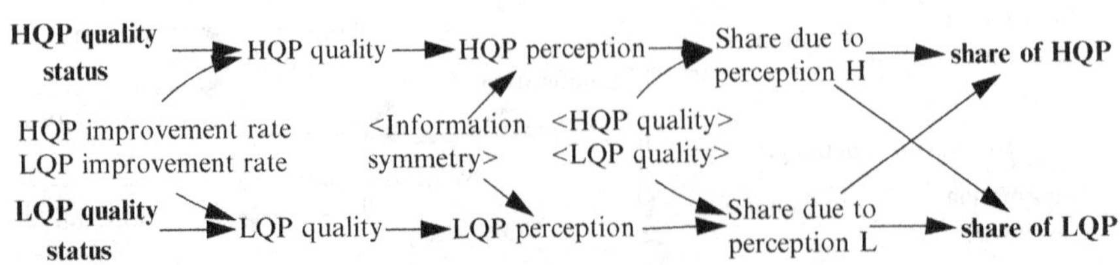

Fig. 6.12 Industry sector

Fig. 6.13 Socioeconomic (poverty) sector

Perception equations are formulated using theories of probability and reliability engineering. Both, information symmetry and product quality are required for probability of quality perception. Market share is determined on the basis of percentage of quality perception. Market share is then segmented on percentage basis into shares of HQP and LQP:

$$HQP_{perception} = \text{Information symmetry} \times \text{HQP quality}$$
$$LQP_{perception} = \text{Information symmetry} \times \text{LQP quality}$$

6.5.8 Socioeconomic (Poverty) Sector

Figure 6.13 shows the application of poverty benefit to LQP. Current poverty status and decline rate of poverty form the input to this sector. Share of HQP is retained whereas poverty effect is applied to the share of LQP. Revised shares are moderated so that their combined share should be 100%. Moderated shares are called HQP and LQP shares. Thus, market share is the final product of the information proliferation, industry, and poverty sectors. It acts as an input to the market sector. In absence of poverty, this sector processes no data, and share of HQP and share of LQP from industry sector are returned as HQP share and LQP share, respectively.

6.6 Observations from Simulation Runs

6.6.1 Process of Quality Perception

Quality perception is a product of quality and information asymmetry. Simulation of quality perception is shown in Fig. 6.14. Information symmetry is increasing linearly, whereas product quality has linear accent and then saturation. Quality perception follows the combined path. Behaviors of information symmetry as well as product quality are seen reflected in the behavior of quality perception.

It is worthwhile to observe the behavior of quality perception along the changing level of information symmetry. Rising reputation and advertising will lead to a rise in information symmetry. Advertising and reputation levels are increased at a time by executing the experimentation plan diagonally. A1, B2, and C3 are chosen for this purpose: A1 for lower, B2 for moderate, and C3 for higher values of information

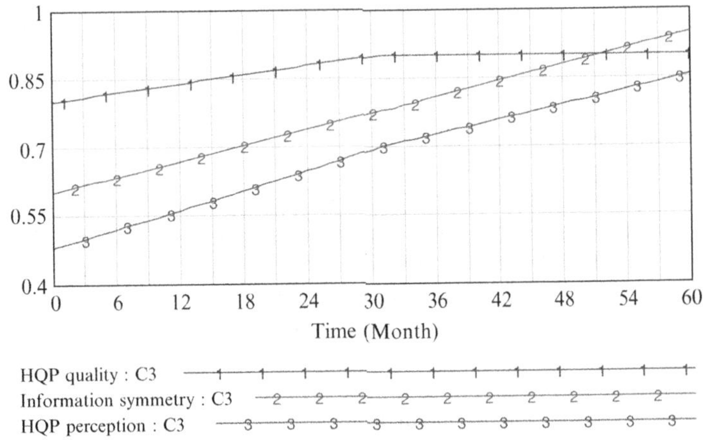

Fig. 6.14 Process of quality perception

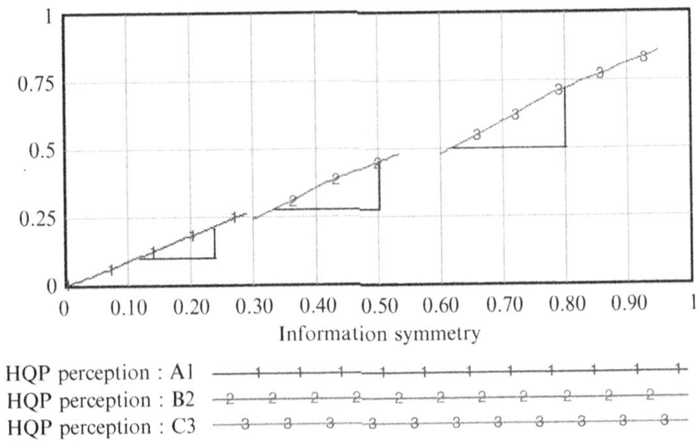

Fig. 6.15 Quality perception patterns

symmetry. Quality perception of HQP is simulated for these cases. Patterns are obtained on a single plot of Fig. 6.15. Rising information symmetry expedites the process of quality perception. Quality perception may behave linearly or may follow a cervical path. Slope measured on the pattern is a rate of quality perception. Rate of quality perception is useful for achieving target quality perception. Rate of quality perception should be determined on the basis of company's competitive requirement of quality perception through the product life cycle. Rate of quality perception helps decide the required advertising drive or mix of instruments to satisfy the necessary information symmetry. Parameter optimization for required quality perception is also feasible.

6.6.2 *Quality Perception in Developed Nations*

Quality perception is different in developed and developing nations. Quality perception in developed nations is considered solely as a result of the presence of information symmetry. A1 and B2 are taken as sample cases from simulation runs for the understanding of quality perception in developed nations. Figure 6.16 shows simulation results for conditions of A1. Reputation is zero and advertising target is lowest for A1. Product quality is significantly high. But level of information symmetry is low. Hence, HQP perception is abysmally low. This makes the market conditions unfavorable to HQP. Little market niche of HQP is shown in Fig. 6.16. LQP gains the position at the cost of HQP.

Figure 6.17 shows advertising and reputation conditions of B2. Moderate conditions lead to moderate information symmetry. Product quality is at the earlier level. Now quality perception is considerable. As shown in Fig. 6.17, LQP niche is still higher than HQP niche. By the end of simulation time, HQP and LQP niches are nearly equal.

Quality perception is proportional to product information symmetry present in the market. Information symmetry is achieved through reputation and advertising drive in our model of simulation study. Thus, reputation and advertising are responsible for quality perception. Reputation and advertising effects on quality perception are shown separately in Fig. 6.18. Simulations are run for B1, B2, B3, C1, C2, and C3. As per the experimentation plan, quality perception gap after simulations of B1 and C1 or B2 and C2 is due to the difference in advertising target. HQP perception difference due to advertising and word-of-mouth is shown in Fig. 6.18. Similarly reputation effect is to be seen from horizontal simulation runs of experimentation plan. Perception change due to reputation is also shown in

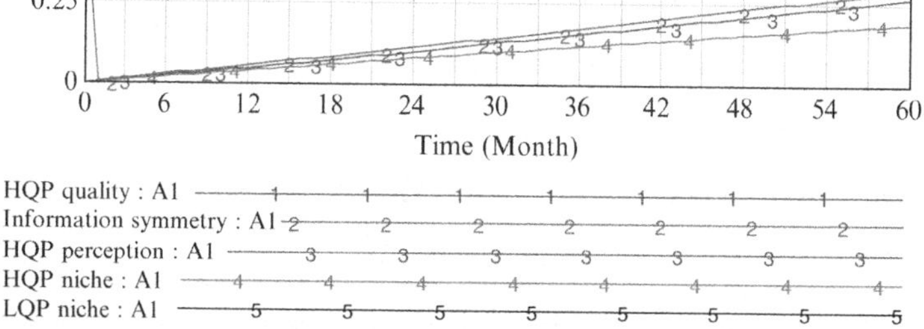

Fig. 6.16 Market parameters for A1

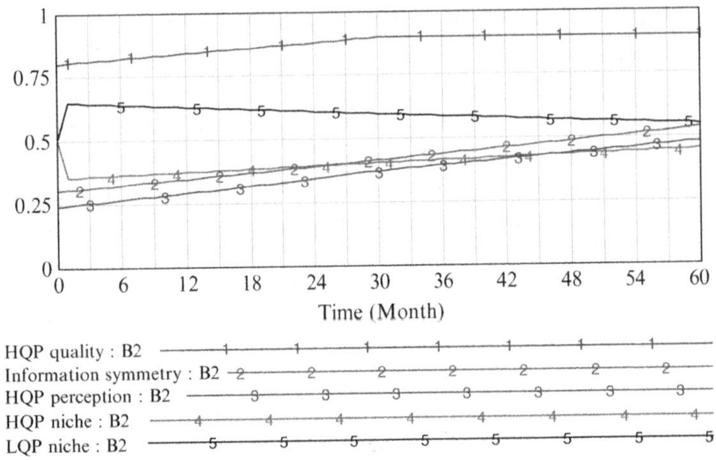

Fig. 6.17 Market parameters for B2

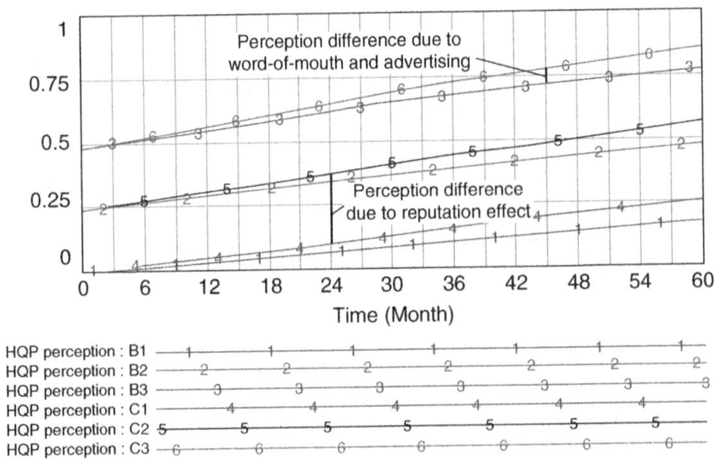

Fig. 6.18 Quality perception due to advertising and reputation

Fig. 6.18. Area of obtained or measured slopes can be used for further study of advertising expenditure or reputation change in the market.

6.6.3 *Quality Perception in Developing Nations*

The phenomenon of quality perception due to information symmetry is more peculiar to developing nations. Quality perception in developing nations can also be ascribed to literacy and amount of purchasing power parity. India is taken as a specific case of developing nations. Figure 6.19 shows the simulation of D1. It is

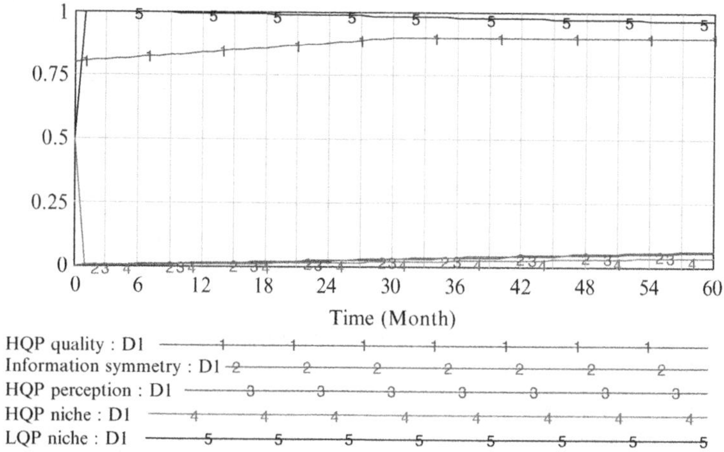

Fig. 6.19 Market parameters for D1

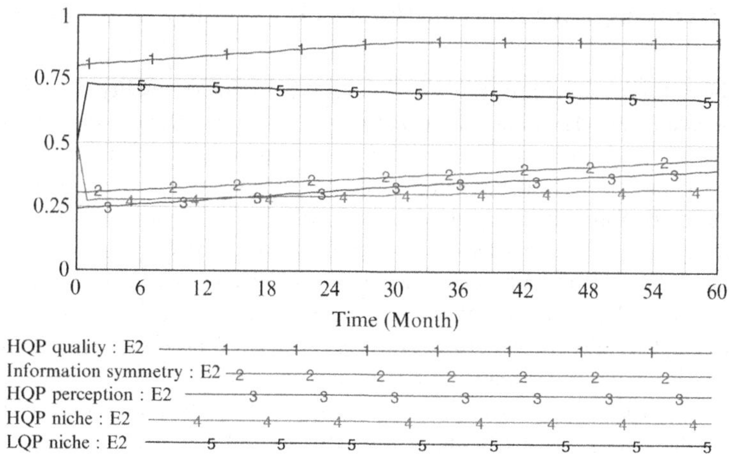

Fig. 6.20 Market parameters for E2

similar to that of A1 except for the application of literacy and purchasing power parity. Information symmetry and hence quality perception are negligible because of these socio-economic causes. This is a specific case where HQPs are driven out of the market by LQPs.

Similarly, simulations for E2 are shown in Fig. 6.20. HQP perception is sizably lower than in the case of B2. Compared to the developed nations' case of B2, wider gap is observed in LQP and HQP niche in Fig. 6.20. Market parameter change shown in Fig. 6.20 is due to the effect of socio-economic conditions of developing nations. This impact is shown in Fig. 6.21. Impact measured in slopes or areas is useful for the further study.

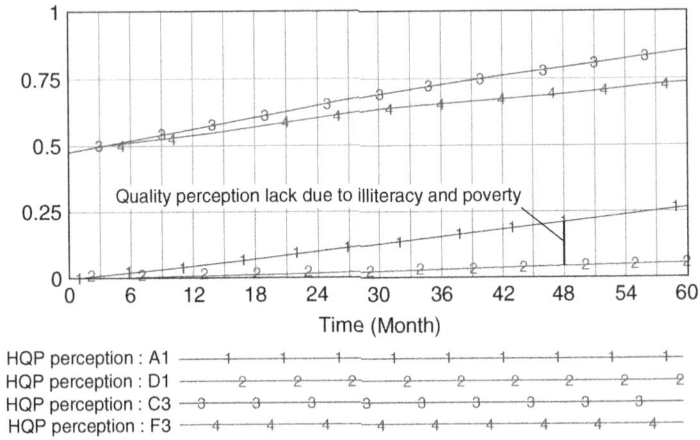

Fig. 6.21 Socio-economic impact on quality perception

6.7 Evolving Hierarchy of Policies for Quality Management

The simulations are helpful to understand the behavior of quality perception and related parameters. The data obtained through the simulation runs, if used properly, make sense for the process of decision-making. Product quality, information symmetry, reputation, advertising, word-of-mouth, and market niche have been considered for the performance measure of quality perception. Impact of these factors on quality perception can be sensed by using the simulated data.

Analytic hierarchy approach (AHP) is a decision-making process when it includes several hierarchies of criteria. Scores are arrived at on the basis of a criterion and weight attached to it. An event with maximum score is selected as a decision. Principle of normalization employed in AHP in its initial steps is found suitable for determining a set of policies. This set of hierarchical policies helps the process of decision making to improve the quality perception. Policies are evolved by exploring the parametric relationships with the help of correlation analysis and subsequent implementation of AHP.

6.7.1 Correlation Performances

Correlation analysis is not meant for unfolding of the causal relationships. Rather it is used to understand the strength of relationships. Product quality, quality perception, information symmetry, reputation, advertising, and market niche are the elements of system dynamics model and hence they are considered pair-wise to measure the correlation performance. A pair of variables under consideration and the resultant data from simulation runs are taken on a excel sheet for each representative case and correlation performances are measured. Correlation values are

Table 6.5 Correlation performances

Parametric relationships	A1	B2	C3	D1	E2	F3
HQP quality and HQP perception	0.891187	0.919499	0.927491	0.891561	0.931099	0.939925
Info. symmetry and HQP perception	0.999926	0.998077	0.996916	0.999926	0.99567	0.993298
HQP quality and HQP niche	0.912289	0.908671	0.915168	0.895902	0.904031	0.911014
Info. symmetry and HQP niche	0.998714	0.999348	0.998799	0.999948	0.999631	0.999074
Literacy and HQP perception	–	–	–	0.999737	0.996773	0.994489
Purchasing power and HQP perception	–	–	–	0.999737	0.996773	0.994489
Literacy and HQP niche	–	–	–	0.999967	0.999867	0.99945
Purchasing power and HQP niche	–	–	–	0.999967	0.999867	0.99945

determined for each pair of variables. Correlation values for these parametric relationships for a general case as well as for developing nations are shown in Table 6.5.

Illustration of an entry in the first row and first column of Table 6.5.

1. The simulations are run for A1 from the experimentation plan.
2. The simulated data sets of HQP quality and HQP perception for this simulation run are pasted on an excel sheet.
3. A correlation value is measured.

Representative combinations A1, B2, and C3 for developed nations and D1, E2, and F3 for developing nations are used for correlation analysis. A value nearer to + 1 suggests stronger relationship and one nearer to −1 indicates a weaker relationship. All the correlation performances indicate strong relations. Some conclusions are drawn on the basis of correlation analysis:

1. The correlation values are increasing significantly from A1 to B2 to C3 as well as from D1 to E2 to F3 for "HQP quality and HQP perception". Increasing values indicate that perception of the quality products will increase by using advertising and reputation.
2. Highest correlation values are obtained for "information symmetry and HQP Perception" for the general case in Table 6.5. Thus, strongest relationship is observed between information symmetry and quality perception. Hence, information symmetry is a market end instrument of quality management.
3. The correlation value of A1 is higher than that of B2 which in turn is higher than that of C3 for "Information symmetry and HQP perception" for the general case in Table 6.5. Similar observations are true for developing nations (D1, E2, and F3). Similarly, value of C3 (F3) is higher than that of B2 (E2) which in turn is higher than that of A1 (D1) for "HQP Quality and HQP perception." This suggests that with the rise of company reputation and advertising drive, the relative role of product quality toward quality perception decreases.

4. Noticeable change in the correlation values of D1, E2, and F3 is not observed for "Literacy and HQP niche." Such observation is also true for "Purchasing power and HQP niche." This indicates strong relationship between literacy and HQP niche as well as between purchasing power and HQP niche for developing nations. This suggests the importance of literacy and purchasing power in developing nations. Carving out the market niche of HQP has inherent limitations in developing nations.

6.7.2 Analytic Hierarchy Approach

Because of multivariate relationships (vertical and horizontal in Table 6.5) the correlation performers are not sufficient to arrive at strategies or policy decisions. Specifically, the aim of simulation runs is to look into the role of reputation and advertising on market parameters. Hence, AHP is a suitable tool to develop managerial implications from this study.

AHP is carried out separately for the general case and developing nations. AHP used here is simply on the basis of the process of normalization as the weight that can be attached is one on parametric as well as experimentation side. First, rows and columns are normalized by averaging the values. Vertical and horizontal normalized scores are then added together for any entity. Table 6.6 summarizes the scores after applying AHP.

Illustration of an entry in the first row and first column of Table 6.6

Step 1: Sum of the column entries from A1 from Table 6.5

$0.891187 + 0.999926 + 0.912289 + 0.998714 = 3.802116$

Step 2: $0.891187/3.802116 = 0.234392$

Step 3: Sum of the row entries from "HQP quality and HQP perception" from Table 6.5.

First three entries, that is, under A1, B2, and C3 are for the general case or developed nations

$0.891187 + 0.919499 + 0.927491 = 2.738177$

Table 6.6 Scores by using AHP

Parametric relationships	A1	B2	C3	D1	E2	F3
HQP quality and HQP perception	0.55986	0.576162	0.580362	0.437224	0.456049	0.460257
Info. symmetry and HQP perception	0.596866	0.594151	0.592593	0.462961	0.460386	0.459168
HQP quality and HQP niche	0.573366	0.569625	0.572902	0.44553	0.449024	0.452382
Info. symmetry and HQP niche	0.595927	0.594692	0.593496	0.461882	0.461129	0.460751
Literacy and HQP perception	–	–	–	0.462638	0.460662	0.459485
Purchasing power and HQP perception	–	–	–	0.462638	0.460662	0.459485
Literacy and HQP niche	–	–	–	0.461821	0.461168	0.460854
Purchasing power and HQP niche	–	–	–	0.461821	0.461168	0.460854

Step 4: 0.891187/2.738177 = 0.325467
Step 5: 0.234392 + 0.325467 = 0.55986

6.7.3 Important Hierarchy of Policies

Policies that are hierarchical in importance can be derived from the AHP scores. These policies serve as guiding principles to improve the quality perception. The scores obtained in Table 6.6 are ranked in descending order. The highest score is selected for determining the top most policy of decision-making. The parameters related to this row and column are explored to set the policy. For example 0.595866 is a highest score after AHP. Corresponding parametric relationship for this score is Information symmetry and HQP perception. On experimentation side, low reputation and low advertising drive correspond to this score. On the basis of these finding, the first policy from Table 6.7 is developed. Similarly, four to five important policies are prepared for developed and developing nations. Important policies for developed nations are shown in Table 6.7 and for developing nations in Table 6.8

Table 6.7 Policy rankings for developed nations

Rank	AHP score	Policy
1	0.596866	Increasing information symmetry by normal advertising for attaining quality perception of HQP
2	0.595927	Increasing information symmetry by normal advertising to carve out maximum market niche of HQP
3	0.594692	Increasing information symmetry by using moderate reputation and advertising drive to carve out the market niche of HQP
4	0.594151	Increasing information symmetry by using moderate reputation and advertising drive for attaining quality perception of HQP

Table 6.8 Policy rankings for developing nations

Rank	AHP score	Policy
1	0.462961	Attaining quality perception of HQP by increasing information symmetry at the normal rate of advertising and reputation
2	0.462638	Attaining quality perception of HQP by increasing literacy rate (beyond the domain of quality management and state has to play the role here)
2	0.462638	Attaining quality perception of HQP by increasing purchasing power parity of the people (beyond the domain of quality management and state has to play the role here)
3	0.461882	Increasing information symmetry by normal advertising to carve out market niche of HQP
4	0.461821	Carving out the market niche of HQP by increasing literacy rate (beyond the domain of quality management and state has to play the role here)
4	0.461821	Carving out the market niche of HQP by increasing purchasing power parity of the people (beyond the domain of quality management and state has to play the role here)

6.8 Managerial Implications

6.8.1 Developed Nations

The policies obtained have managerial implications of significance. Findings from simulation and analysis emphasize the role of information symmetry for increasing quality perception and market niche of HQP in developed nations. Hence, information symmetry has to play a key role for quality perception in developed nations. Also normal mode of advertising is indicated by policy rankings. This suggests that advertising should be oriented toward information content.

6.8.2 Developing Nations

Developing nations have inherent limitations in raising the quality perception. Policy rankings ascribe the key role of literacy and purchasing power to quality perception in developing nations. Still, the role of increased information symmetry at the normal rate of advertising and reputation for quality perception of HQP is the most desirable outcome in the case of developing nations. Also, possible use of advertising for quality perception and to carve out the market niche of HQP is shown by another arrived policy.

Policies observed for developing nations need specific attention. The literature on information economics considers information asymmetry as the most responsible parameter for quality uncertainty in developing nations. But correlation analysis and policies from using AHP strongly suggest the role of literacy and purchasing

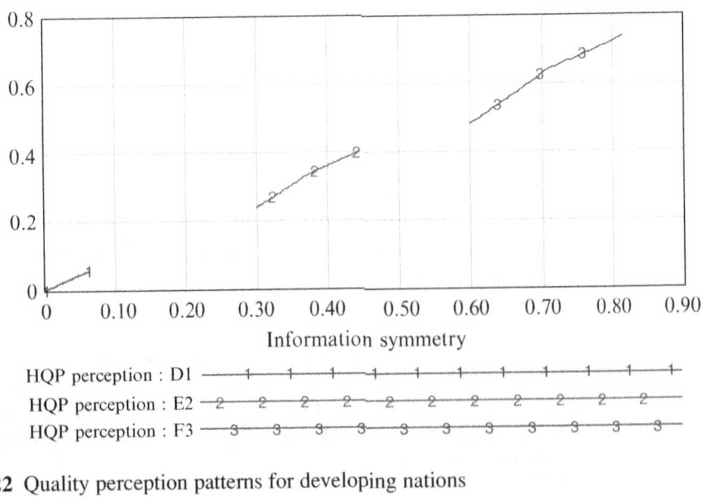

Fig. 6.22 Quality perception patterns for developing nations

power. These revelations are significant for the fact that these modify earlier theories. Information symmetry definitely plays a role in quality perception in developing nations. But at the lower values of information symmetry, the process of quality perception is slow and hinders further course of quality perception. These facts are also supported by the HQP perception patterns obtained in Fig. 6.22.

Academically, the real market is largely an intersection of economics, quality management, and marketing science. Market with information symmetry, quality perception, and other factors is a complex and dynamic one. All inclusive theoretical background to quality perception is developed in this chapter. Simulations are important to theory building when issues are complex. Even preliminary modeling and simulation runs are sufficient to provide an insight into the issue. Simulation and analysis done here is a beginning of the systems analysis of the market. It may offer a long term foundation to quality perception and market dynamics. Attempts have to be made to incorporate all possible factors. It may not be possible to crystallize a few factors to their fullest extent. But they should be considered to the sufficient extent for making market dynamics nearer to reality.

Appendix C

Worksheets 5 and 6

Using Worksheet 5

Worksheet 5 refers to the first part of the system dynamics modeling and analysis of quality perception. It helps develop causal loop and boundary chart for the modeling. Entries in worksheet 5 are as follows:

1. Imagine the basic sectors – company, market, information to customers, etc. Ponder over the TQM efforts the company has been investing in. Also, imagine the marketing strategies the company has been developing and the current scenario of all these. Now draw a possible causal loop diagram for quality perception.
2. The four sectors are identified in the column 1. These sectors are considered separately for simplicity and clarity. The socio-economic sector can be neglected if not required.
3. Make a list of the endogenous factors for all the sectors in the second column. These are the variables for system dynamics modeling that are internal parts of the system. The four sectors are the four subsystems for this system dynamics modeling.
4. Make a list of the exogenous factors for all the sectors in the third column. These are the variables that are external parts of the system. These influence the system from outside.
5. Make a list of the excluded factors for all the sectors in the third column. These are the variables that are least important from modeling viewpoint at the first iteration.

Using Worksheet 6

Worksheet 6 refers to the system dynamics modeling of quality perception. The modeling work has been broken down in the four submodels. These submodels should be appropriately connected to one another by using the shadow variables. Entries in worksheet 6 are as follows:

1. Draw a model for the industry or company sector in the first row. The efforts for improvement of product quality are modeled here. TQM implementation is a representation of the product quality. Hence, the critical factors should be modeled. The other important aspects should also be considered.
2. The second row should present the market position of the company. Reputation, market share, and other performance measures determine the modeling efforts in this row.

3. The modeling in the third row is a kernel of the modeling of quality perception. All the attempts by the company to increase information symmetry of product quality have to be incorporated in this model. Or this is a model where one can run the simulations and determine the amount of required marketing enablers for the desired quality perception.
4. The modeling in the fourth row largely depends on whether the company is placed in developed or developing nations and on the background of the company's customers.

While developing the models lots of inputs are required for the second and the third models. Thus, a company team should do the research and develop these models, whereas the inputs are readily available with the company for the first model. Continuous research and iterations are essential for this modeling. The model developed on the quarter page should grow to a page size model after iterations.

Worksheet 5

Causal loop diagram

Boundary chart

Sector	Endogenous	Exogenous	Excluded
Company/ industry			
Market			
Information			
Socio- economic			

Worksheet 6

Sector	Model
Company This sector should incorporate TQM and other improvement activities at the company level	
Market This sector should position the company juxtaposed to the competitors. Reputation or market share is important	
Information Marketing strategies or enablers like advertising, warranty, word of mouth, etc. should become the kernel of this sector	
Socio-economic The impact of socio-economic and cultural factors on quality perception should be presented here	

Chapter 7
Future Directions of Quality Perception

Growing importance has been attached to quality perception. Garvin (1984) described quality perception as a dimension of quality management. Then, Brown (1995) coined the term "nation equity" for perceived quality of a product from an individual nation's point of view. Contemporarily, marketing people look at it from customers' point of view as a sample survey and assessment. During a similar period, quality uncertainty was being developed as a measure of bad quality, and the business studies attempted metrics that could be linked to customers' loyalty.

However, quality management had a few traces on it. Recent literature on quality management has found that there is an absence of a universally accepted TQM model so far and more studies are required to understand the quality paradigm. Picking the thread from this, the concept of quality perception has been developed in this book in a multidisciplinary perspective. Initially, the phenomenon of quality uncertainty is explored conceptually. Without substantial foundation, it could have been a hasty attempt toward empiricism. So, many pages of this book are aimed at the theoretical foundation of quality perception. Exploring in any single direction may or may not necessarily yield a suitable metric, as the research topic of quality perception is naïve per se to a multidisciplinary domain. The corroboration of what is arrived at might become another issue of concern. So, multidirectional approach to the problem has been attempted.

Thus, the work on quality perception has taken the path as shown in Fig. 7.1. The prevailing literature on information economics, marketing science, business studies, and quality management has revealed the commonality and the gaps on quality perception. Hence, the literature on these disciplines has been the basis for the work on quality perception. Basic theory building process that links all these disciplines is completed in this book. Also, sufficient instruments, to contain quality uncertainty or to promote quality perception, are formulated. The following tasks (from 7.1 to 7.7) on quality perception are accomplished in this book.

L. Wankhade and B. Dabade, *Quality Uncertainty and Perception*,
Contributions to Management Science, DOI 10.1007/978-3-7908-2195-6_7,

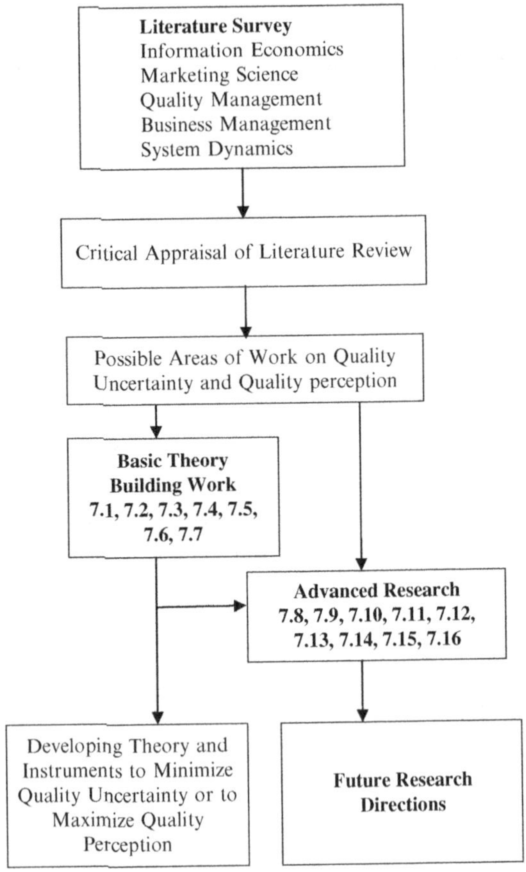

Fig. 7.1 Quality perception accomplishments and future directions

7.1 Illustration of Quality Uncertainty

Quality uncertainty is illustrated in the literature on information economics. The basis of such illustrations is economics viewpoint. There was a need of quality uncertainty illustration with respect to product quality. Now, quality uncertainty has been explained in the manufacturing parlance. We are able to understand quality uncertainty with respect to cost, profit, and market niche of any product.

Now, the purview of quality management has been expanded to marketing. While making the policies on TQM, it has become essential that a company has to see its reflection in a market. Equally, the role of lacking marketing parameters to cause quality uncertainty has surfaced to prominence.

7.2 System Thinking of Quality Uncertainty

Systems analysis is a suitable tool to crystallize the complex issue of quality uncertainty. Systems thinking and analysis have been used to identify the causes of information asymmetry. An attempt is made to delve into information asymmetry and quality uncertainty. A causal loop exhibiting the causes of quality uncertainty or quality perception is drawn. The causal loop represents the basic sectors responsible for quality uncertainty or perception. The causal loop can be translated into a generic model of quality perception.

7.3 Cause and Effect Analysis of Quality Uncertainty

Quality uncertainty is a dynamic process where causes other than information asymmetry are possible. Hence, root cause analysis is carried out to understand the causes and effects of quality uncertainty. Some of the factors involved in the process are subjective. The cause and effect analysis is found effective in this situation. Every factor is an independent entity in cause and effect analysis. Hence, it is possible to know the role of any individual factor in generating quality uncertainty. Also, the management people are conversant with the seven management tools which are used to determine the extent of quality uncertainty and to plan the priorities to minimize it.

7.4 Probabilistic Modeling of Quality Uncertainty

Quality uncertainty contains the term "uncertainty" clearly indicating the possible use of theory of probability. Also the term "uncertainty" suggests that it is a failure of quality to translate into the market value. Mathematical model of quality uncertainty due to information asymmetry is developed by using the concepts of probability and reliability engineering. Probabilistic modeling rendered a more scientific treatment to quality uncertainty than any kind of approximation.

7.5 Remodeling of Expected Quality or Perceived Quality

Because of the presence of information asymmetry, customers value high-quality product as an average of high and low. This averaging is called as "expected quality" in information economics. In the presence of marketing parameters like advertising, reputation of the company, guarantee, warranty, etc., the "objective quality" becomes "perceived quality" in marketing science literature. Thus a need emerged to remodel "expected quality" or "perceived quality" in the backdrop of probabilistic modeling of quality uncertainty. Also, a relationship has been found between quality uncertainty and quality perception. The remodeling and the reemerged relationships have attempted to bridge the existing gap in the literature for quality uncertainty, perceived quality, and quality management.

7.6 Behavioral Study of Quality Perception Through Simulations

The system dynamics models are developed for quality perception. A few parameters are measurable and a few are immeasurable. Simulations provide a beautiful stratagem in such a situation. The process of quality perception has been crystallized. Quality perception has been analyzed for developed and developing nations. The role of governing parameters has been exhibited, and policy rankings are also developed. The modeling and simulation procedure that has been developed can be extended toward further work on quality perception.

7.7 Theory Building of Quality Uncertainty and Perception

A careful review of the current research in various disciplines reveals that emerging quality dimensions in all these disciplines point toward a direction that seems similar but different in content. The approaches followed in these disciplines have been incomplete as they are pursued mostly in isolation of each other. Thus, for quality perception, quality management has been assimilated into marketing science by using the connection of information economics. Assimilation of these three disciplines has evolved a sound theory of quality perception.

Instruments to contain quality uncertainty or to enhance quality perception have been suggested. The beauty is that the instrument can be selected on the basis of the difficulty level. The approximate method of root cause analysis is a lower level approach. This is more appealing as it uses the seven management tools. The business specific factors which develop quality uncertainty should be collected as a group study. Then develop an interrelationship matrix and carry out the further analysis. Working step by step upon the priorities that are generated by using the root cause instrument minimizes quality uncertainty.

The mathematical approach is a middle level approach. This method requires the knowledge of the theories of probability and reliability engineering. Again, collect the different sets of factors specific to a business house. Do the computation. If the factors are large enough in the fault tree diagram, then employ a cut set. Enhance upon the market enablers and recompute quality perception. Reiterate the procedure to gain higher level of quality perception.

System dynamics based approach is the most precise and higher level approach to maximize quality perception. It also uses the formulations from the middle level mathematical approach. Also, all possible subtleties can be used while developing a system dynamics model. Simulations help foresee the improved business performance. This method requires knowledge of system dynamics apart from the theories of probability. System dynamics modeling of quality perception is particularly important as it can translate every piece of information into modeling.

The mathematical model, the framework for the market dynamics using quality perception, and the method of root cause analysis are ready for various business applications. Although the tools developed in this book follow the scientific method of enquiry to make their validity easier, further refining is still possible. Hence, following avenues (from 7.8 to 7.16) are opened for future work on quality perception.

7.8 Refining the Dimensions of Quality Perception

The dimensions of quality perception have been evolved through literature and the process of theory building. Akerlof (1970) identified the factors that stem quality uncertainty. The quality uncertainty factors have been translated into the quality perception factors while quality perception is coined positively. Further, TQM constructs produce quality at origin. These constructs are responsible to make the quality either an uncertain or a perceived one at the first place. Hence, these constructs have become a part of the quality perception process.

The set of factors causing information asymmetry and thereby quality uncertainty has two subsets. One of the subsets describes the socio-economic and cultural conditions that cause quality uncertainty. These include a few measurable and a few immeasurable items. The other subset enumerates the marketing enablers that enhance quality perception. At this juncture, more marketing strategy factors have been incorporated into the process.

Thus, many dimensions play a role in quality perception and the fact is that still these dimensions need further exploration. While some dimensions are purely independent of each other, a few are overlapping to some extent. For example, reputation is an outcome of many performance measures including quality perception. However, initial reputation of the company can be an independent dimension of quality perception. The degree of overlap is a matter of enquiry. And further work is required to identify the overlap and to purify or refine the dimensions by coefficient alpha (Cronbach 1951) method or an appropriate statistical method.

7.9 Parameters' Optimization for Quality Perception

Various market enablers like guarantee, warranty, rebate, and advertising are used to enhance quality perception. These are discussed in the literature for impact measurement or strategy formulations. Mostly, one or two factors find a place as a part of any assessment.

It has been observed that parametric efforts at a market end are unorganized and are not linked to the delivered results. This makes improper revenue utilization of a company resulting in competitive disadvantage. Quality perception as a strategy of business performance needs all these parameters considered at a place. Hence, the market enablers should be optimized for the desired quality perception. Optimization and the desired business improvement are possible by using mathematical or system dynamics approach to quality perception.

Another application of product quality perception model is its use in categorizing a firm's customers into several perceived-quality segments (e.g., high, medium, and low). These segments then can be analyzed on the basis of (Parasuraman et al. 1988) (1) demographic and/or other profiles; (2) the relative importance of individual dimension in influencing product quality perception; and (3) the reasons behind the perceptions reported. On the basis of this analysis, dimensions' optimization should be practiced to fulfil the set targets.

7.10 Modeling of Product Information Proliferation Rate

Product information proliferation rate of the preliminary system dynamics model of quality perception is a key but fuzzy parameter with socio-economic complexity. Though proliferation rate is a complex phenomenon, its modeling nearer to a reality is certainly possible.

Intensive field explorations are required here. The first will be a nation specific or a region specific one which will be on the basis of the first subset of the factors of information asymmetry. This subset has socio-economic and cultural studies and it will be applicable to any business from the region or nation concerned.

Second subset specifically pertains to the marketing enablers used by a company. This subset is relatively easier for the modeling task than the first one. Elaborate system dynamics model for information proliferation rate should be attempted where both these subsets should work in tandem.

7.11 Study of Information Asymmetry in Supply Chain Processes

Product quality is received in a market through supply chain processes. It is worthwhile to investigate the presence of information asymmetry at various points in the supply chain processes. Quality uncertainty or perception will be a process of continuous accumulation at every point in a supply chain. Thus, total quality perception at a market end will be determined on the basis of the products' flow in the supply chain.

7.12 TQM Constructs and Quality Perception

The level of TQM implementation in any industry depends on the level of implementation of its constructs. TQM constructs at industry end are broken down into tangible performance measures. System dynamics model of TQMI can be developed to study the effect of various measures on TQMI as well as quality perception.

7.13 Product Life Cycle and Quality Perception

Product life cycle and quality perception have symbiotic relationship. If quality perception has been growing at the natural pace in its life cycle, then perhaps word-of-mouth will be the only enabler of quality perception. The quality perception will be increasing arbitrarily and the fate will be largely determined by the competitive players and their efforts to enhance quality perception of their products.

Hence, expenditure toward market enablers and the points in a life cycle where market enablers are employed at varying proportions decide the level of quality perception in the life cycle. If quality perception of particular product grows along with market growth of product type, then it is likely that company offering it will garner a business volume equal to that of predicted market niche. The lack or delayed rise in quality perception will cause commensurate revenue loss to the company. Further studies are required to understand the relationship between product life cycle and quality perception in the backdrop of product lifetime revenue.

7.14 Industry Specific Modeling of Quality Perception

A particular industry would like to know the quality perception of its products to decide management strategies. A firm should take into account any distortion of customers' perceptions of product quality and see that a reference standard of quality is established to decide what product quality the firm will like to have (Ali and Seshadri 1993).

Both, marketing and quality managers believed that marketing should play an important role in ensuring that quality strategies reflect "real" customer attitudes and beliefs by collecting and analyzing market information and actively representing the "voice of the customer" in quality strategy formulation. Many marketers and a smaller number of quality managers also suggested a role for marketing in the implementation of quality strategy in terms of marketing communication and its potential impact on customers' subjective quality perceptions (Morgan and Piercy 1998).

The general system dynamics model then deems industry/company specific reorientation. For this task, industry specific performance measures are important. Product life cycle, TQMI, working of supply chain, reputation, and enablers like advertising, warranty, guarantee, and rebate are industry specific performers inducing quality perception. The detailed system dynamics model for an industry specific case can be formulated on the basis of endeavors toward product quality improvement (TQMI) and marketing efforts.

7.15 Development of a Robust Measure of Quality Perception

There is a need to incorporate concepts from marketing science, like carryover effect of quality, memory retention time of product quality, and advertising frequency to further advance the model of quality perception. Similarly, perceived quality relies upon advertisements since quality is a subjective characteristic and "false advertising" of quality is difficult to prove (Ali and Seshadri 1993).

It is observed that quality perception is also based on the product type conditions. For example, quality perception experiences for durable products can be different from those of nondurable products. For instance, consider comparing computer with candy bar – in the case of candy bars, the importance of externalities, signaling, and psychological status associated with either exclusivity or popularity is low. For personal computers, all positive network externalities (availability of software as well as after-sale services) are relevant (Hellofs and Jacobson 1999).

7.16 Studying the Impact of Perceived Quality on Stock Market

Aaker and Jacobson (1994) have studied the effect of perceived quality on stock market. Perceived quality was considered then as an intangible entity. They observed two opposing views on whether a firm should supply product quality information. Current-term performance measures (e.g., Return On Investment (ROI)) will not reflect the long term effects of several marketing variables. The observed association between stock return and perceived product quality should be encouraging to those attempting to justify investments in product quality. The stock market, once given reliable indicators of this information, will react and rely less on short-term measures of business performance (Aaker and Jacobson 1994). Now, as a universal model is being developed, quality perception is becoming a firm and reliable entity representing TQM constructs as well as marketing parameters. It is worthwhile to study the validity of quality perception for short-term and long-term repercussions on stock market.

References

Aaker DA, Jacobson R (1994) The financial information content of perceived quality. J Mark Res 31(2):191–201

Abernethy AM, Butler DD (1992) Advertising information: service versus product. J Retailing 68 (4):398–419

Ahire S, Golhar D, Waller M (1996) Development and validation of TQM implementation constructs. Decis Sci 27(1):23–56

Akao, Yoji (1966) Development history of quality function deployment. The customer driven approach to quality planning and deployment. Minato, Tokyo 107 Japan: Asian Productivity Organization

Akerlof G (1970) The market for lemons: quality uncertainty and market mechanism. Q J Econ 84 (3):485–500

Akerlof G (2001) Behavioral microeconomics and macroeconomic behaviour. Nobel Prize Lecture, pp 1–30

Akerlof G (2003) Writing the "The market for 'lemons'": A personal and interpretive essay. http://www.nobelprize.org/economics/articles/akerlof/

Ali A, Seshadri S (1993) Customer perception and competitive quality strategy. Managerial Decis Econ 14(3):235–246

Allen F (1984) Reputation and product quality. RAND J Econ 15(3):311–327

Andersson EW (1998) Customer satisfaction and word of mouth. J Serv Res 1(1):5–17

Andrews IR, Valenzi VR (1971) Combining price, brand name, and store cues to form an impression of product quality. Proceedings of the 79th annual convention of the American Psychological Association, pp 649–650

Arrow K (1963) Uncertainty and welfare economics of medical care. Am Econ Rev 58:941–973

Badri AM, Davis D (1995) A study of measuring the critical factors of quality management. Int J Qual Reliab Manage 12(2):36–53

Bass F (1969) A new product growth model for consumer durables. Manage Sci 15(5):215–227

Bass F, Krishnan T, Jain D (1994) Why the bass model fits without decision variables. Mark Sci 13 (3):203–223

Besterfield DH, Besterfield-Michna C, Besterfield GH, Besterfield-Sacre M (2003) Total quality management. Pearson Education, Singapore

Black S, Porter L (1996) Identification of the critical factors of TQM. Decis Sci 27(1):1–21

Bolton RN, Lemon KN (1999) A dynamic model of customer's usage of services: usage as an antecedent and consequence of satisfaction. J Mark Res 36(2):171–186

Boulding W, Kalra A, Staelin R (1999) The quality double whammy. Mark Sci 18(4):463–484

Boulding W, Kalra A, Staelin R, Zeithaml V (1993) A dynamic process model of service quality: from expectations to behavioral intentions. J Mark Res 30(1):7–27

Bounds G, Lyle Y, Mel A, Gipsie R (1994) Beyond total quality management: towards the emerging paradigm. McGraw-Hill, New York

Brady MK, Cronin JJ (2001) Some new thoughts on conceptualizing perceived service quality: a hierarchical approach. J Mark 65:34–49

Brown J, Reingen PH (1987) Social ties and word-of-mouth referral behaviour. J Consum Res 14:350–362

Brown JA (1995) Perceptions of quality in the global market. TQM Mag 7(6):52–56

Carman JM (1990) Consumer perceptions of service quality: an assessment of the SERVQUAL dimensions. J Retailing 66(1):33–55

Carter D, Baker B (1992) Concurrent engineering: the product developing environment for the 1990s. Addison-Wesley, Reading, MA

Connor P, Newton D, Bromley R (2002) Practical reliability engineering. Wiley, England

Cravens DW, Holland CW, Lamb CW, Moncrief WC (1988) Marketing's role in product and service quality. Ind Mark Manage 17:285–304

Cronbach LJ (1951) Coefficient alpha and internal structure of tests. Psychometrika 16 (10):297–334

Cronin J, Taylor S (1992) Measuring service quality: a reexamination and extension. J Mark 56:55–88

Crosby PB (1979) Quality is free. New American Library, New York, NY

Crosby PB (1989a) Quality management: something more than super quality control. Exec Excell 6(5):13–14

Crosby PB (1989b) How goes the quality revolution? J Qual Participation 12(1):83–88

Curry A, Kkolou E (2004) Evaluating CRM to contribute to TQM improvement – a cross-case comparison. TQM Mag 16(5):314–324

Dabade BM, Ray PK (1996) Quality engineering for continuous performance improvement in products and processes: a review and reflections. Qual Reliab Eng Int 12:173–189

Dahalgaard JJ, Kristensen K, Kanji GK, Juhl HJ, Sohal AS (1998) Quality management practices: a comparative study between East and West. Int J Qual Reliab Manage 15(8&9):812–826

Dale B, McQuater R (1998) Managing business improvement and quality: implementing key tools and techniques. Blackwell Business, Oxford

Davis JP, Bingham CB, Eisenhardt KM (2007) Developing theory through simulation methods. Acad Manage Rev 32(2):580–599

Day G (1994) The capabilities of market-driven organizations. J Mark 58(4):37–52

Deming WE (1981) Improvement of quality and productivity through action by management. Natl Productivity Rev 1(1):12–22

Deming WE (1982) Quality productivity and competitive position. MIT Center for Advanced Engineering, Cambridge, MA

Deming WE (1986) Out of the crisis. MIT Center for Advanced Engineering, Cambridge, MA

Dodds WB, Monroe KB (1985) The effect of brand and price information on subjective product evaluation. Adv Consum Res XII:85–90

Doggett AM (2004) A statistical comparison of three root cause analysis tools. J Ind Technol 20 (2):2–9

Dybvig PH, Spatt CS (1983) Does it pay to maintain a reputation? Consumer information and product quality. Yale University, Mimeo

Feigenbaum AV (1961) Total quality control. McGraw-Hill, New York, NY

Feigenbaum AV (1983) Total quality control: engineering and management. McGraw-Hill, New York, NY

Flynn B, Schoeder R, Sakibaba S (1994) A framework for quality management research and associated measurement instrument. J Oper Manage 11:339–366

Fombrun CJ (1996) Reputation: realizing value from the corporate image. Harvard Business School Press, Boston, MA

Forrester JW (1961) Industrial dynamics. MIT Press, Cambridge, MA

Forrester JW (1968) Principles of systems. Pegasus Communications, Waltham, MA

Forrester JW (1969) Urban dynamics. Pegasus Communications, Waltham, MA
Forrester JW (1971) World dynamics. Pegasus Communications, Waltham, MA
Forrester JW (1995) Beginning of system dynamics. Mckinsey Q 4:4–16
Fudenberg D, Kreps DM (1987) Reputation in the simultaneous play of multiple opponents. Rev Econ Stud 54(4):541–568
Fudenberg D, Levine D (1992) Maintaining a reputation when strategies are imperfectly observed. Rev Econ Stud 59(3):561–579
Gale B (1994) Managing customer value. Simon & Schuster, New York, NY
Garvin DA (1983) Quality on the line. Harv Bus Rev 61(5):65–75
Garvin D (1984) What does product quality really means? Sloan Manage Rev 26(1):25–43
Gonzalez ME, Quesada G, Mueller R (2004) QFD strategy house: an innovative tool for linking marketing and manufacturing strategies. Mark Intell Plan 22(3):335–348
Greising D (1994) Quality: how to make it pay. Business Week, 8 Aug, pp 54–59
Gronroos C (1982) Strategic management and marketing in the service sector. Swedish School of Economics and Business administration, Helsingfors
Gronroos C (1984) A service quality model and its marketing implications. Eur J Mark 18 (4):36–44
Gupta AK (1996) Reliability engineering and terotechnology. Macmillan India Limited, New Delhi
Hellofs LL, Jacobson R (1999) Market share and customers' perceptions of quality: when can firms grow their way to higher versus lower quality? J Mark 63(1):16–25
Hjorth-Anderson C (1984) The concepts of quality and the efficiency of markets for consumer products. J Consum Res 11(9):708–718
Hollbrook MB, Corfman KP (1985) Quality and value in the consumption experience: Phaldrus rides again. In: Jacoby J, Olson J (eds) Perceived quality. Lexington Books, Lexington, MA, pp 31–57
Horner J (1999) Reputation and competition. CARESS Working Paper, 99–02
Ishikawa K (1976) Guide to quality control. Asian Productivity Organization, Tokyo
Ishikawa K (1988) Group wide quality control. J Qual Participation 11(1):4–6
Ismail S, Maling E (2002) An investigation of the total quality management survey based research published between 1989 and 2000: a literature review. Int J Qual Reliab Manage 19(7):902–970
Iyer G, Soberman D, Villas-Boas JM (2005) The targeting of advertising. Mark Sci 24(3):461–476
Jabnoun N (2000) Restructuring for TQM: a review. TQM Mag 12(6):395–399
Jacob R (1993) TQM: more than a dying fad. Fortune 18:66–72
Jacoby J, Olson J, Haddock RA (1971) Price, brand name and product composition characteristics as determinants of product quality. J Appl Psychol 55(6):570–579
Jacoby J, Olson J (eds) (1985) Perceived quality. Lexington Books, Lexington, MA
Jagadeesh R (1999) Total quality management in India – perspective and analysis. TQM Mag 11 (5):321–327
Jermain E (1963) Problems in industrial dynamic. MIT Press, Cambridge, MA
Joseph IN, Rajendran C, Kamalanabhan TJ (1999) An instrument for measuring total quality management implementation in manufacturing – based business units in India. Int J Prod Res 37(10):2201–2215
Juran JM (1974) Quality control handbook, 3rd edn. McGraw-Hill, New York, NY
Juran JM (1986) The quality trilogy. Qual Prog 9(8):19–24
Juran JM (1988) Managing for quality. J Qual Participation 11(1):8–12
Kaplan RS, Norton DP (1992) The balanced scorecard – measures that drive performance. Harvard Business Review 70(1):71–78
Keiningham TL, Vavra T (2001) The customer delight principle. McGraw-Hill, New York
Keiningham TL, Aksoy L, Cooil B, Andreassen TW (2008) Linking customer loyalty to growth. MIT Sloan Manage Rev 49(4):51–57
Khanna VK, Vrat P, Shankar R, Sahay BS (2004) Managing transition phases in the TQM journey: a system dynamics approach. Int J Qual Reliab Manage 21(5):518–544

Kirmani A (1990) The effect of perceived advertising costs on brand perceptions. J Consum Res 17:160–171
Kirmani A (1997) Advertising repetition as a signal of quality: if it is advertised so much, something must be wrong. J Advert 26(3):77–86
Kopalle PK, Lehmann DR (1995) The effects of advertised and observed quality on expectations about new product quality. J Mark Res 32(3):280–290
Kopalle PK, Lehman DR (2006) Setting quality expectations when entering a market: what should the promise be? Mark Sci 25(1):8–24
Llusar JCB, Zornoza CC (2002) Development and validation of a perceived business quality measurement instrument. Qual Manage J 9(4):23–38
McAdam R, Henderson J (2004) Influencing the future of TQM: internal and external driving factors. Int J Qual Reliab Manage 21(1):51–71
Mitra D, Golder PN (2006) How does objective quality affect perceived quality? Short-term effects, long-term effects, and asymmetries. Mark Sci 25(3):230–247
Moorman C (1998) Market-level effects of information: competitive responses and consumer dynamics. J Mark Res 35(2):82–98
Moorman C, Du R, Mela C (2005) The effect of standardized information on firm survival and marketing strategies. Mark Sci 24(2):263–274
Moorthy S, Zhao H (2000) Advertising spending and perceived quality. Mark Lett 11(3):221–233
Morgan NA, Piercy NF (1998) Interactions between marketing and quality at the SBU level: influences and outcomes. J Acad Mark Sci 26(3):190–208
Motwani J (2001) Critical factors and performance measures of TQM. TQM Mag 13(4):292–300
Nerlove M (1958) Adaptive expectations and cobweb phenomena. Q J Econ 73:227–240
Noronha C (2003) National culture and total quality management: empirical assessment of a theoretical model. TQM Mag 15(5):351–355
Olshavsky RW, Miller JA (1972) Consumer expectations, product performance, and perceived product quality. J Mark Res 9(1):19–21
Olson JC, Jacoby J (1972) Cue utilization in the quality perception process. Third annual conference of the association for consumer research association for consumer research, Iowa City
Olson JC (1977) Price as an informational cue: effects in product evaluation. In: Woodside A, Sheth J, Bennet P (eds) Consumer and industrial buying behavior. North Holland, New York
O'Neal CR, Lafief WC (1992) Marketing's lead role in total quality. Ind Mark Manage 21:133–143
Parasuraman A, Zeithaml V, Berry L (1985) A conceptual model of service quality and its implications for future research. J Mark 49:41–50
Parasuraman A, Zeithaml V, Berry L (1988) SERVQUAL: a multiple-item scale for measuring consumer perceptions of service quality. J Retailing 64(1):12–40
Park C-H, Kim Y-G (2003) A framework of dynamic CRM: linking marketing with information strategy. Bus Proc Manage J 9:652–671
Phillips LW, Chang DR, Buzzell RD (1983) Product quality, cost position and business performance: a test of some key hypotheses. J Mark Res 47:26–43
Polly L (1994) Quality's silent partner. Industry Week, 18 April, 47–48
Powell TC (1995) Total quality management as competitive advantage: a review and empirical study. Strateg Manage Study 16:15–37
Prasad S, Tata J (2003) The role of socio-cultural, political-legal, economic, and educational dimensions in quality management. Int J Oper Prod Manage 23(5):487–521
Reichheld FF (2003) The one number you need to grow. Harv Bus Rev 81(12):46–54
Reingen PH, Kernan JB (1986) Analysis of referral networks in marketing: methods and illustrations. J Mark Res 23:370–378
Rogerson WP (1983) Reputation and product quality. Bell J Econ 14:508–516
Rosen DL, Olshavsky RW (1987) A protocol analysis of brand choice strategies involving recommendations. J Consum Res 14:440–444

Royal Swedish Academy of Sciences (2001) Markets with asymmetric Information. Royal Swedish Academy of Sciences, Stockholm

Saraph J, Benson P, Schroeder R (1989) An instrument for measuring the critical factors of quality management. Decis Sci 20:810–829

Shapiro C (1982) Consumer information, product quality, and seller reputation. Bell J Econ 13 (1):20–35

Shapiro C (1983) Premiums for high quality products as returns to reputation. Q J Econ 98 (4):659–679

Sice P, Mosekilde E, Moscardini A, Lawler K, French I (2000) Using system dynamics to analyze interactions in duopoly competition. Syst Dyn Rev 16(2):113–133

Sjolander R (1992) Cross-cultural effects of price on perceived product quality. Eur J Mark 26 (7):34–44

Scitovszky T (1945) Some consequences of the habit of judging quality by price. Rev Econ Stud 12:100–105

Solomon MR (1986) The missing link: surrogate consumers in the marketing chain. J Mark 50:208–218

Spence M (2001) Signalling in retrospect and the informational structure of the market. Nobel Prize Lecture, pp 1–30

Steenkamp JB (1990) Conceptual model of the quality perception process. J Bus Res 21 (4):309–333

Sterman JD (2000) Business dynamics: systems thinking and modeling for a complex world. Irwin/McGraw-Hill, New York, NY

Stiglitz J (2000) The contributions of the economics of information to twentieth century economics. Q J Econ 115(4):1441–1478

Stiglitz J (2001) Information and the change in the paradigm in economics. Nobel Prize Lecture, pp 1–69

Tar JJ (2005) Components of successful total quality management. TQM Mag 17(2):182–194

Tellis GJ, Wernerfelt B (1987) Competitive price and quality and under asymmetric information. Mark Sci 6:240–253

Tellis GJ, Fornell C (1988) The relationship between advertising and product quality over the product life cycle: a contingency theory. J Mark Res 25(1):64–71

Tellis GJ (1988) Advertising exposure, loyalty, and brand purchase: a two-stage model of choice. J Mark Res 25(2):134–144

Tellis GJ (1989) The impact of corporate size and strategy on competitive pricing. Strateg Manage J 10(6):569–585

Tellis GJ, Prabhu J (2000) Do consumers ever learn? Analysis of segment behavior in experimental markets. J Behav Decis Mak 3(1):19–34

Tellis GJ, Chandy RK, MacInnis D, Thaivanich P (2005) Modeling the microeffects of television advertising, which ad works, when, where, for how long, and why? Mark Sci 24(3):351–366

Vojtko V, Mildeova S, Trojasek J, Nemecova I (2006) Market dynamics for decision support in marketing. 24th International conference of the System Dynamics Society, System Dynamics Society, Nijmejen, Netherlands

Wankhade L, Dabade BM (2004) Information asymmetry and quality perception. Proceedings of the second international conference on logistic and supply chain management, Coimbatore, India

Wankhade L, Dabade BM (2005) A holistic approach to quality success in developing nations. TQM Mag 17(4):322–328

Wankhade L, Dabade BM (2006a) Analysis of quality uncertainty due to information asymmetry. Int J Qual Reliab Manage 23(2):230–241

Wankhade L, Dabade BM (2006b) TQM with quality perception: a system dynamics approach. TQM Mag 18(4):341–357

Wankhade L, Dabade BM (2007a) Perceived quality as an emerging dimension of quality management. Proceedings of the 8th international research conference on quality, innovation and knowledge management, New Delhi, India, 12–14 Feb 2007

Wankhade L, Dabade BM (2007b) Information symmetry, quality perception, and market dynamics. Int J Modell Manage 2(3):208–231

Wankhade L, Dabade BM (2009) Minimising quality uncertainty by the root-cause and failure analysis method. Int J Manage Decis Mak 10(5/6):359–385

Weick KE (1989) Theory construction as disciplined imagination. Acad Manage Rev 14 (4):516–531

Zeithaml V (1987) Defining and relating price, perceived quality, and perceived value, Report No. 87-101. Marketing Science Institute, Cambridge, MA

Zeitz G, Johannesson R, Ritchie JE Jr (1997) An employee survey measuring total quality management practices and culture. Group Organ Manage 22(4):414–444

About the Authors

Lalit Wankhade, Ph.D., is a Professor of Production Engineering at SGGS Institute of Engineering and Technology, Nanded, India. He teaches Operations Management, Reliability Engineering, and System Dynamics. He received his Master of Engineering in Computer-Aided Design and Computer-Aided Manufacturing from SGGS Institute of Engineering and Technology, Nanded. He received his Ph.D. from SRTM University, Nanded, India. He has research interests in information asymmetry, product quality, and system dynamics. He has introduced the topic of quality uncertainty to quality management. He has modeled quality perception by using the theories of reliability engineering and developed a system dynamics framework for quality perception. He has been engaged in research on modeling of quality perception.

Balaji Dabade, Ph.D., is a Professor of Production Engineering at SGGS Institute of Engineering and Technology, Nanded, India. He received his Ph.D. from Indian Institute of Technology, Kharagpur, India. He has been teaching Quality Engineering at the college level for more than 20 years. He has published widely. His papers have been published in *International Journal of Quality Engineering*, *International Journal of Quality and Reliability Management*, *International Journal of Production Research*, *The TQM Magazine*, etc. His current teaching and research interests include quality engineering, design of experiments, optimization techniques, and total quality management.

Index

Made in the USA
Monee, IL
18 October 2026

66037579R10085